Basic Trigonometry Functions

$$\sin\theta = \frac{\text{opposite side}}{\text{hypotenuse}} = \frac{O}{H} \qquad -90° \le \left[\theta = \sin^{-1}\left(\frac{O}{H}\right)\right] \le 90°$$

$$\cos\theta = \frac{\text{adjacent side}}{\text{hypotenuse}} = \frac{A}{H} \qquad 0° \le \left[\theta = \cos^{-1}\left(\frac{A}{H}\right)\right] \le 180°$$

$$\tan\theta = \frac{\text{opposite side}}{\text{adjacent side}} = \frac{O}{A} \qquad -90° \le \left[\theta = \tan^{-1}\left(\frac{O}{A}\right)\right] \le 90°$$

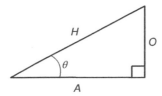

Sine Law

$$\frac{A}{\sin\theta_A} = \frac{B}{\sin\theta_B} = \frac{C}{\sin\theta_C}$$

Cosine Law

$$C^2 = A^2 + B^2 - 2AB\cos\theta_C$$

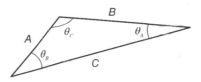

$$\sin(-\theta) = -\sin(\theta) \qquad\qquad \sin(\theta + 90°) = \cos(\theta)$$

$$\cos(-\theta) = \cos(\theta) \qquad\qquad \sin(\theta - 90°) = -\cos(\theta)$$

$$\sin(\theta + \phi) = \cos\theta\sin\phi + \sin\theta\cos\phi \qquad \sin(\theta \pm 180°) = -\sin(\theta)$$

$$\sin(\theta - \phi) = \cos\theta\sin\phi - \sin\theta\cos\phi \qquad \cos(\theta + 90°) = -\sin(\theta)$$

$$\cos(\theta + \phi) = \cos\theta\cos\phi - \sin\theta\sin\phi \qquad \cos(\theta - 90°) = \sin(\theta)$$

$$\cos(\theta - \phi) = \cos\theta\cos\phi + \sin\theta\sin\phi \qquad \cos(\theta \pm 180°) = -\cos(\theta)$$

DESIGN AND ANALYSIS OF MECHANISMS

DESIGN AND ANALYSIS OF MECHANISMS
A PLANAR APPROACH

Michael J. Rider, Ph.D.
Professor of Mechanical Engineering, Ohio Northern University, USA

This edition first published 2015
© 2015 John Wiley & Sons, Ltd.

Registered Office

John Wiley & Sons, Ltd, The Atrium, Southern Gate, Chichester, West Sussex, PO19 8SQ, United Kingdom.

For details of our global editorial offices, for customer services and for information about how to apply for permission to reuse the copyright material in this book please see our website at www.wiley.com.

Library of Congress Cataloging-in-Publication Data

Rider, Michael J.
Design and analysis of mechanisms : a planar approach / Michael J. Rider, Ph.D.
    pages   cm
Includes bibliographical references and index.
ISBN 978-1-119-05433-7 (pbk.)
1. Gearing. 2. Mechanical movements. I. Title.
TJ181.R53 2015
621.8′15–dc23
                                        2015004426

A catalogue record for this book is available from the British Library.

Set in 10/12pt Times by SPi Global, Pondicherry, India

1   2015

Contents

Preface

The intent of this book is to provide a teaching tool that features a straightforward presentation of basic principles while having the rigor to serve as basis for more advanced work. This text is meant to be used in a single-semester course, which introduces the basics of planar mechanisms. Advanced topics are not covered in this text because the semester time frame does not allow these advanced topics to be covered. Although the book is intended as a textbook, it has been written so that it can also serve as a reference book for planar mechanism kinematics. This is a topic of fundamental importance to mechanical engineers.

Chapter 1 contains sections on basic kinematics of planar linkages, calculating the degrees of freedom, looking at inversions, and checking the assembling of planar linkages. Chapter 2 looks at position analysis, both graphical and analytical, along with a vector approach, which is the author's preferred method. Chapter 3 looks at graphical design of planar linkages including four-bar linkages, slider–crank mechanisms, and six-bar linkages. Chapter 4 looks at the analytical design of the same planar linkages found in the previous chapter. Chapter 5 deals with velocity analysis of planar linkages including the relative velocity method, the instant center method, and the vector approach. Chapter 6 deals with the acceleration analysis of planar linkages including the relative acceleration method and the vector approach. Chapter 7 deals with the static force analysis of planar linkages including free body diagrams, equations for static equilibrium, and solving a system of linear equations. Chapter 8 deals with the dynamic force analysis based on Newton's law of motion, conservation of energy and conservation of momentum. Adding a flywheel to the mechanism is also investigated in this chapter. Chapter 9 deals with spur gears, contact ratios, interference, basic gear equations, simple gear trains, compound gear trains, and planetary gear trains. Chapter 10 deals with fundamental cam design while looking at different types of followers and different types of follower motion and determining the cam's profile.

There are numerous problems at the end of each chapter to test the student's understanding of the subject matter.

Appendix A discusses the basics of using the Engineering Equation Solver (EES) and how it can be used to solve planar mechanism problems. Appendix B discusses the basics of MATLAB and how it can be used to solve planar mechanism problems.

1

Introduction to Mechanisms

1.1 Introduction

Engineering involves the design and analysis of machines that deal with the conversion of energy from one source to another using the basic principles of science. Solid mechanics is one of these branches. It contains three major sub-branches: kinematics, statics, and kinetics. Kinematics deals with the study of relative motion. Statics is the study of forces and moments apart from motion. Kinetics deals with the result of forces and moments on bodies. The combination of kinematics and kinetics is referred to as dynamics. However, dynamics deals with the study of motion caused by forces and torques. For mechanism design, the desired motion is known and the task is to determine the type of mechanism along with the required forces and torques to produce the desired motion. This text covers some of the mathematics, kinematics, and kinetics required to perform planar mechanism design and analysis.

A mechanism is a mechanical device that transfers motion and/or force from a source to an output. A linkage consists of links generally considered rigid which are connected by joints such as pins or sliders. A kinematic chain with at least one fixed link becomes a mechanism if at least two other links can move. Since linkages make up simple mechanisms and can be designed to perform complex tasks, they are discussed throughout this book.

A large majority of mechanisms exhibit motion such that all the links move in parallel planes. This text emphasizes this type of motion, which is called two-dimensional planar motion. Planar rigid body motion consists of rotation about an axis perpendicular to the plane of motion and translation in the plane of motion. For this text, all links are assumed rigid bodies.

Mechanisms are used in a variety of machines and devices. The simplest closed form linkage is a 4-bar, which has three moving links plus one fixed link and four pinned joints. The link that does not move is called the ground link. The link that is connected to the power source is called the input link. The follower link contains a moving pivot point relative to ground and it is typically considered as

Design and Analysis of Mechanisms: A Planar Approach, First Edition. Michael J. Rider.
© 2015 John Wiley & Sons, Ltd. Published 2015 by John Wiley & Sons, Ltd.

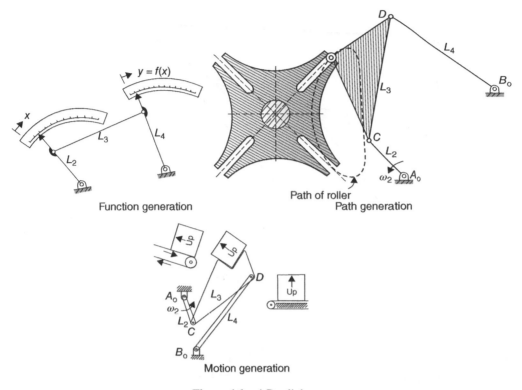

Figure 1.1   4-Bar linkages

the output link. The coupler link consists of two moving pivots, points C and D, thereby coupling the input link to the output link. A point on the coupler link generally traces out a sixth-order algebraic coupler curve. Very different coupler curves can be generated by using a different tracer point on the coupler link. Hrones and Nelson's *Analysis of 4-Bar Linkages* [1] published in 1951 shows many different types of coupler curves and their appropriate 4-bar linkage.

The 4-bar linkage is the most common chain of pin-connected links that allows relative motion between the links (see Figure 1.1). These linkages can be classified into three categories depending on the task that the linkage performs: function generation, path generation, and motion generation. A **function generator** is a linkage in which the relative motion or forces between the links connected to ground is of interest. In function generation, the task does not require a tracer point on the coupler link. In **path generation**, only the path of the tracer point on the coupler link is important and not the rotation of the coupler link. In **motion generation**, the entire motion of the coupler link is important, that is, the path that the tracer point follows and the angular orientation of the coupler link.

1.2   Kinematic Diagrams

The first step in designing or analyzing a mechanical linkage is to draw the kinematic diagram. A kinematic diagram is a "stick-figure" representation of the linkage as shown in Figure 1.2.

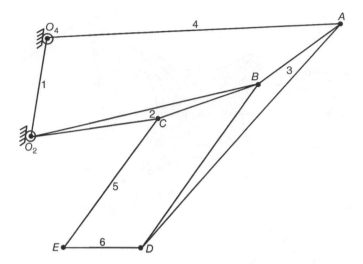

Figure 1.2 Kinematic diagram

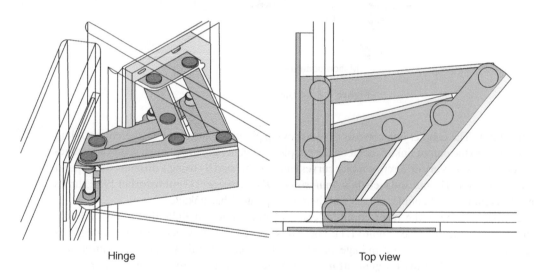

Hinge   Top view

Figure 1.3 Physical system

The kinematic diagram is made up of nodes and straight lines and serves the same purpose as an electrical circuit schematic used for design and analysis purposes. It is a simplified version of the system so you can concentrate on the analysis and design instead of the building of the system. The actual 3D model is shown in Figure 1.3.

For convenience, the links are numbered starting with the ground link as number 1, the input link as number 2, then proceeding through the linkage. The purpose of a kinematic diagram is to show the relative motion between links. For example, a slider depicts translation while a pin joint depicts rotation. The joints are lettered starting with letter A, B, C, etc. On some kinematic

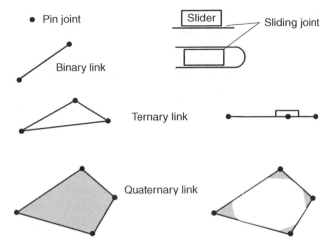

Figure 1.4   Planar links and joints

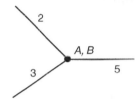

Figure 1.5   Two joints where three links join

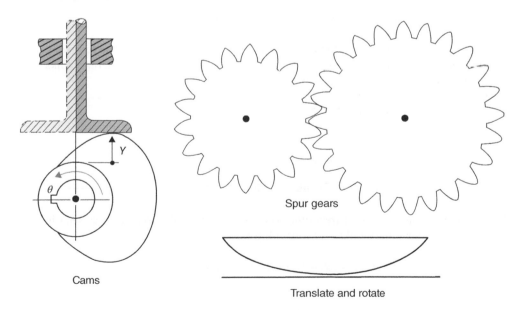

Figure 1.6   Half joints

diagrams, it is preferred to label fixed rotational pin joints using the letter O; thus link 2 connected to ground at a fixed bearing would be labeled O_2 and link 4 connected to ground at a fixed bearing would be O_4. Both notations are used in this book.

A link is a rigid body with at least two nodes. A node is a point on a link that attaches to another link. Connecting two links together forms a joint. The two most common types of nodes are the pin joint and the sliding joint; each has one degree of freedom. Links are categorized by the number of joints present on them. For example, a binary link has two nodes and a ternary link has three nodes (see Figure 1.4).

If three links come together at a point, the point must be considered as two joints since a joint is the connection between two links, not three links (see Figure 1.5).

A full joint has one degree of freedom. A half joint has two degree of freedom. Figure 1.6 shows half joints which can translate and rotate. A system with one degree of freedom requires one input to move all links. A system with two degrees of freedom requires two inputs to move all links. Thus, the degrees of freedom represent the required number of inputs for a given system.

1.3   Degrees of Freedom or Mobility

Kutzbach's criterion for 2D planar linkages calculates the number of degrees of freedom or mobility for a given linkage.

$$M = 3(L-1) - 2J_1 - J_2$$

L = Number of links including ground

J_1 = Number of one degree of freedom joints (full joints)

J_2 = Number of two degrees of freedom joints (half joints)

If we consider only full joints, then the mobility can also be calculated using the following equation which is a modification of Gruebler's equation. Note that the number of ternary links in the mechanism does not affect its mobility.

$$M = B - Q - 2P - 3$$

B = Number of binary links (2 nodes)

Q = Number of quaternary links (4 nodes)

P = Number of pentagonal links (5 nodes)

A 4-bar linkage has one degree of freedom. So does a slider-crank mechanism as seen in Figure 1.7. Each has four binary links and four full joints.

$$M = 3(L-1) - 2J_1 - J_2 = 3(4-1) - 2(4) - 0 = 1$$

or

$$M = B - Q - 2P - 3 = 4 - 0 - 2(0) - 3 = 1$$

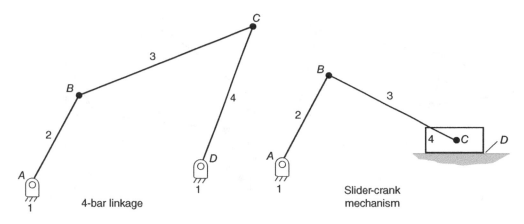

Figure 1.7   Mechanisms with four links

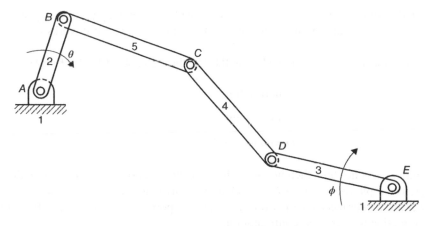

Figure 1.8   Five-bar linkage

A 5-bar linkage has two degrees of freedom. It has five binary links and five full joints. The 5-bar requires two separate inputs as shown in Figure 1.8.

$$M = 3(5-1) - 2(5) - 0 = 2$$

A 6-bar linkage with two ternary links has one degree of freedom. It has six links and seven full joints. A 6-bar linkage can be put together in two different configurations, the Watt 6-bar and the Stephenson 6-bar, as shown in Figure 1.9. What is the difference between these configurations?

The Watt 6-bar linkage has the two ternary links connected together, whereas the Stephenson 6-bar linkage has the two ternary link separated by a binary link.

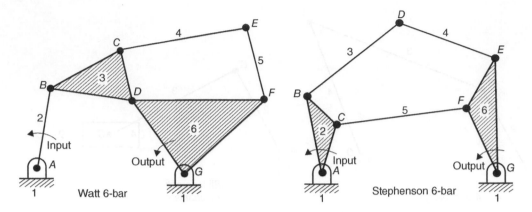

Figure 1.9 6-Bar linkage (2 configurations)

1.4 Grashof's Equation

A mark of a "good design" is simplicity. A design with the fewest moving parts is in general less expensive and more reliable. With this in mind, a 4-bar linkage is best if it works for your application.

Grashof's equation states that at least one link will rotate through 360° if $S + L \leq P + Q$ where

$$S = \text{Shortest link}$$

$$L = \text{Longest link}$$

$$P, Q = \text{Other two links}$$

If Grashof's equation is not true, then no link will rotate through 360°. If $S + L > P + Q$, then the 4-bar linkage is a triple rocker. If $S + L = P + Q$, then all inversions are either double-crank mechanisms or crank-rocker mechanisms with a change-over point where it will move from an open loop configuration to a crossed configuration.

If $S + L \leq P + Q$, then based on where the shortest link is located, the 4-bar linkage is as follows:

a. Crank-crank if S is the ground link
b. Crank-rocker if S is the input link
c. Double-rocker if S is the coupler link (see Figure 1.10)
d. Rocker-crank if S is the output link

1.5 Transmission Angle

In Figure 1.11, the acute angle between the coupler link, L_3, and the follower link, L_4, is called the transmission angle, μ. For equilibrium of link 4, the sum of the torques about point B_o must be zero. Since the coupler link, link 3, is a two-force member, the force that link 3 applies to link 4 is along link 3. Thus, the torque that link 4 experiences about B_o is $\text{Torque}_4 = \text{Force}_{34} L_4 \sin \mu$.

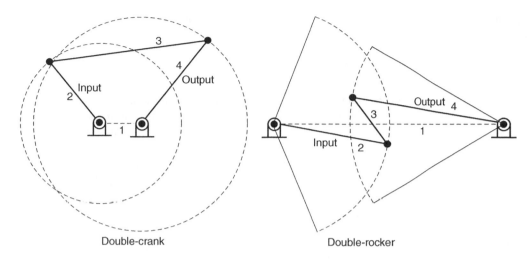

Double-crank   Double-rocker

Figure 1.10   Double-crank and double-rocker

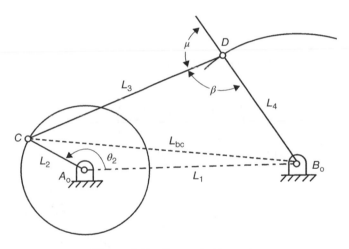

Figure 1.11   Transmission angle

Now looking at the two triangles that have L_{bc} in common, we can use the cosine law and relate θ_2 to β.

$$L_{bc}^2 = L_1^2 + L_2^2 - 2L_1 L_2 \cos\theta_2$$

and

$$L_{bc}^2 = L_3^2 + L_4^2 - 2L_3 L_4 \cos\beta$$

If we set the two equations equal to each other and solve for β, we have:

$$\beta = \cos^{-1}\left(\frac{L_3^2 + L_4^2 - \left(L_1^2 + L_2^2 - 2L_1 L_2 \cos\theta_2\right)}{2L_3 L_4}\right)$$

Since the transmission angle is always an angle less than $90°$, the transmission angle becomes as follows:

$$\mu = \text{Minimum}(\beta, 180° - \beta)$$

A small transmission angle is undesirable for several reasons. As the transmission angle decreases, the output torque on the follower decreases for the same coupler-link force. If the output torque is constant, then the coupler link force must increase as the transmission angle decreases. This could lead to links buckling or connecting pins shearing. Also, as the transmission angle decreases, the position of the follower link becomes more sensitive to linkage lengths and hole tolerances at the connecting pins. To avoid this, the transmission angle should be above $40°$ at all times. Note that $\sin(40°) = 0.64$, and thus the follower torque will be reduced to approximately two-third of its maximum, which occurs when the transmission angle is at $90°$.

For the linkage in Figure 1.12, the extreme values of the transmission angle, μ' and μ'', can be obtained when the input link 2 is aligned with the ground link, link 1.

Using the cosine law again leads to the following:

$$\beta' = \cos^{-1}\left(\frac{L_3^2 + L_4^2 - (L_1 - L_2)^2}{2L_3L_4}\right)$$

$$\mu' = \text{Minimum}(\beta', 180° - \beta')$$

and

$$\beta'' = \cos^{-1}\left(\frac{L_3^2 + L_4^2 - (L_1 + L_2)^2}{2L_3L_4}\right)$$

$$\mu'' = \text{Minimum}(\beta'', 180° - \beta'')$$

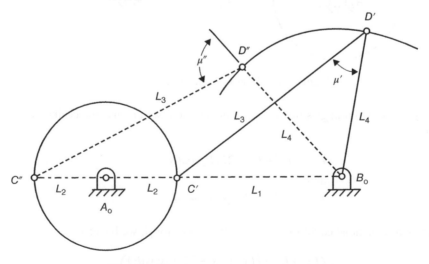

Figure 1.12   Transmission angle extremes

1.6 Geneva Mechanism

A Geneva mechanism converts a constant rotational motion into an intermittent translational motion. The *Geneva Drive* is also called the *Maltese Cross*. The Geneva mechanism was originally invented by a watch maker from Geneva to prevent the spring of a watch from being over-wound. In operation, a drive wheel with a pin enters into one of several slots on the driven wheel and thus advances it by one step. The drive wheel has a raised circular disc that serves to lock the driven wheel in a fixed position between steps. Figure 1.13 shows a five-slot and six-slot Geneva wheel mechanism.

Figure 1.14 shows several shots of the motion of the output wheel as the input wheel rotates.

The following are equations for designing a Geneva wheel mechanism. The first four variables are design parameters chosen by the designer. The rest of the variables are calculated based on these design parameters (see Figure 1.15).

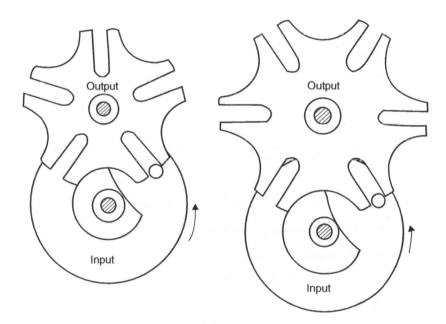

Figure 1.13 Geneva mechanism

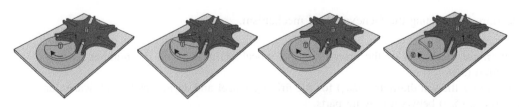

Figure 1.14 Motion of Geneva mechanism

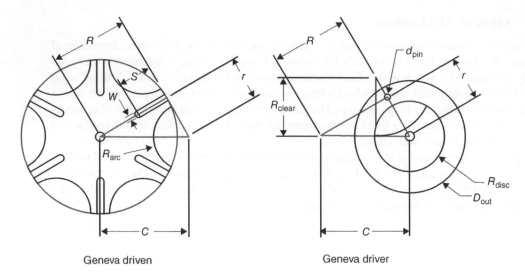

Geneva driven   Geneva driver

Figure 1.15   Geneva wheel design

R = Geneva wheel radius

N = Number of slots

d_{pin} = Diameter of pin

δ_{tol} = Allowed clearance

C = Center distance to drive wheel = $\dfrac{R}{\cos\left(\dfrac{180°}{N}\right)}$

r = Drive crank radius = $R\tan\left(\dfrac{180°}{N}\right) = \sqrt{C^2 - R^2}$

S = Slot center distance = $R + r - C$

W = Slot width = $d_{pin} + \delta_{tol}$

R_{arc} = Stop-arc radius = $r - 1.5 d_{pin}$

R_{disc} = Stop-disc radius = $R_{arc} - \delta_{tol}$

R_{clear} = Clearance-arc radius = $\dfrac{R \cdot R_{disc}}{r}$

D_{out} = Outer diameter of driver = $2r + 3 d_{pin}$

Procedure for creating the Geneva wheel mechanism.

1. Determine the radius of the driven Geneva wheel (R) along with the number of slots required (N).
2. Determine the pin diameter (d_{pin}) for the driving wheel and an acceptable allowance or clearance (δ_{tol}) between moving parts.
3. Calculate the center distance (C) between the driven Geneva wheel and the driver.

4. Calculate the drive crank radius (r) for the driver.
5. Layout triangle C–R–r (see Figure 1.15).
6. Calculate the slot length (S) and width (W).
7. On side (R) of the triangle, layout the slot length and width.
8. Calculate the stop-arc radius (R_{arc}).
9. Draw the stop-arc radius arc centered at the driver location so that it intersects the Geneva wheel radius as shown in Figure 1.15.
10. Using the slot and the arc cutout as a group, pattern it around the driven wheel's center location N times. It should now look like the leftmost picture in Figure 1.15.
11. From the center of the driver wheel, draw the stop-disc radius (R_{disc}).
12. Calculate the clearance arc (R_{clear}) and then draw it perpendicular to the center distance (C) as shown on the right in Figure 1.15. Draw a line parallel with triangle side "r" and through the rotation point of the driver wheel. It should pass through the R_{clear} line just drawn.
13. Draw an arc of radius R_{clear} so that it intersects stop-disc radius circle.
14. Locate the pin diameter (d_{pin}) at a distance of "r" from the rotation point.
15. Calculate the outer diameter of the driver wheel and then add it to the drawing.
16. Clean up the design drawing.

Problems

For Problems 1 through 4, number all links and label all joints.

1.1   Draw the kinematic diagram for the mechanism shown in Figure 1.16. Determine the degrees of freedom for this linkage.

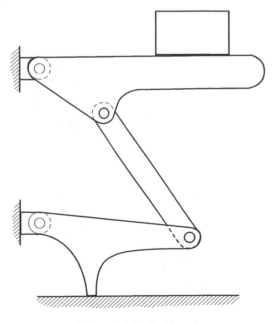

Figure 1.16   Problem 1

1.2   Draw the kinematic diagram for the mechanism shown in Figure 1.17. Determine the degrees of freedom for this linkage.

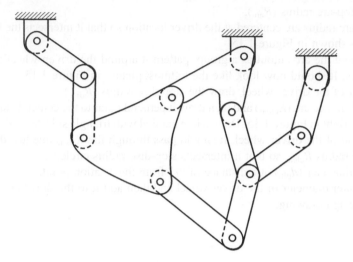

Figure 1.17   Problem 2

1.3   Draw the kinematic diagram for the mechanism shown in Figure 1.18. Determine the degrees of freedom for this linkage.

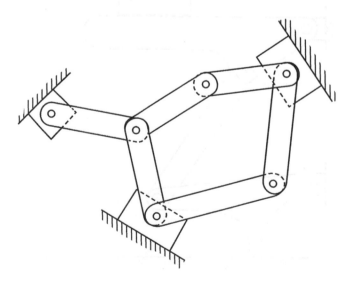

Figure 1.18   Problem 3

1.4   Draw the kinematic diagram for the mechanism shown in Figure 1.19. Determine the degrees of freedom for this linkage.

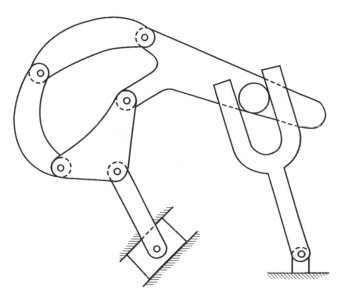

Figure 1.19   Problem 4

1.5   Design a Geneva wheel with four slots. Driven wheel is 6 in. in diameter. Pin diameter is 0.25 in. Clearance is 0.03 in.
1.6   Design a Geneva wheel with five slots. Driven wheel is 150 mm in diameter. Pin diameter is 6 mm. Clearance is 0.8 mm.
1.7   Determine the mobility for the mechanism in Figure 1.20.

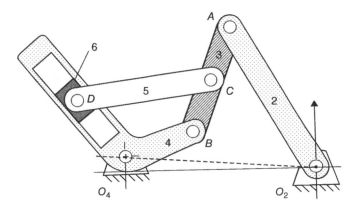

Figure 1.20   Problem 7

1.8   Determine the mobility for the mechanism in Figure 1.21.

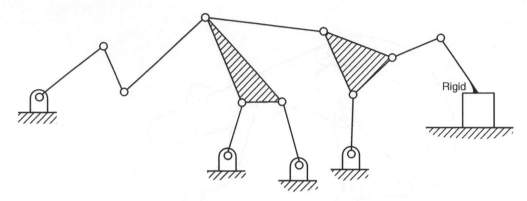

Figure 1.21   Problem 8

Reference

[1] Hrones, John A., and Nelson, George L., *Analysis of the Four-Bar Linkage*, New York: The Technology Press of
     MIT and John Wiley & Sons, Inc. 1951.

2

Position Analysis of Planar Linkages

2.1 Introduction

In analyzing the motion of a planar linkage, the most basic issue encountered is defining the concept of position and displacement. Because motion can be thought of as a time series of displacements between successive positions, it is important to understand the meaning of the term position.

Position is a term that tells where an item is located if the item is a point. Its position can be specified by its distance from a predefined origin and its direction with respect to a set reference axes. If we choose to work with a Cartesian coordinate system, we can specify the position of a point by giving its X and Y coordinates. If we choose to work in a polar coordinate system, then we need to specify the distance from the origin and the angle relative to one of its reference axes. In any case, the position of a point in two-dimensional (2D) space is a vector quantity.

If we want to specify the position of a rigid body, it is necessary to specify more than just its (x, y) coordinates. It is necessary to specify enough information that the location of every point on the rigid body is uniquely determined. A rigid body in 2D space can be defined by two points, A and B, on the rigid body. The position of point B with respect to point A is equal to the position of point B minus the position of point A. Another way to say this is the position of point B can be defined by defining the position of point A and the position of point B relative to point A. Since the object is rigid, all of its points are defined relative to these two points.

Since the purpose of a planar linkage is to move one of its links through a specified motion, or have a point on one of its links move through a specified motion, it is important to be able to verify that the linkage performs its desired function. Thus, position analysis will be covered first.

Successive positions of a moving point define a curve. The curve has no thickness; however, the curve has length because it occupies different positions at different times. This curve is called a path or locus of moving points relative to a predefined coordinate system.

Design and Analysis of Mechanisms: A Planar Approach, First Edition. Michael J. Rider.
© 2015 John Wiley & Sons, Ltd. Published 2015 by John Wiley & Sons, Ltd.

2.2 Graphical Position Analysis

Graphical position analysis can be used quickly to check the location of a point or the orientation of a link for a given input position. All that is needed is a straight edge, a scale, and a protractor. If a parametric CAD system like Creo Parametric® is used, then the user simply needs to sketch the linkage, adjust the sizes and orientation of the known links, then request the position and orientation of the unknown links and/or points. Graphical position analysis is also very useful in checking the analytical position analysis solution at several points to verify that the analytical solution is valid. The graphical solution procedures for 4-bar and slider-crank linkages are outlined below. A similar procedure can be used to draw other linkages such as 6-bars.

2.2.1 Graphical Position Analysis for a 4-Bar

Assume we want to determine the proper orientation for the 4-bar linkage shown in Figure 2.1 when the input, link 2, is at 40°. Note that the input link, L_2, is on the right in this figure.

Given: The origin is located at bearing A_o. Bearing B_o is 3.50 in. left of bearing A_o. The input, link 2, is 1.25 in. long. Link 3 is 4.00 in. long. Link 4 is 2.00 in. long. Distance from point C to point P is 3.00 in. Distance from point D to point P is 2.00 in. What is the angular orientation of links 3 and 4? What is the (x, y) location of point P relative to the origin?

Procedure:

1. Locate the origin and label it A_o. Select an appropriate scale such as 1 model unit = 1 in. (actual size) or 1 model unit = 2 in. (double size).
2. Locate bearing B_o relative to A_o.

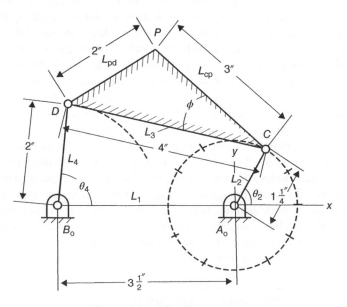

Figure 2.1   4-Bar linkage sketch

3. Draw a line starting at A_o at a 40° angle above the x-axis, then mark off it length of 1.25 model units. Mark this point C.
4. Set your compass at the length of link 4, 2.00 model units. Draw an arc centered at bearing B_o.
5. Set your compass at the length of link 3, 4.00 model units. Draw an arc centered at point C so that it crosses the previously drawn arc. The intersection of these two arcs is point D. (Note there are two intersection points; one above the x-axis for the uncrossed linkage and one below the x-axis for the crossed linkage. Choose the intersection point above the x-axis.)
6. Draw in links 3 and 4.
7. Set your compass at the length of L_{cp}, 3.00 model units. Draw an arc centered at point C.
8. Set your compass at the length of L_{pd}, 2.00 model units. Draw an arc centered at point D so it intersects the previously drawn arc. The intersection of these two arcs is point P. (Note there are two intersection points; choose the proper intersection point.)
9. Draw in the sides, L_{cp} and L_{pd}, to complete the 4-bar linkage.
10. Using a protractor, measure the angular orientation of links 3 and 4.
11. Using your scale, measure the horizontal and vertical distances from the origin to point P.
12. Box in your answers with the appropriate units.

From the graphical solution shown in Figure 2.2, it can be seen that $\theta_3 = -16.0°$, $\theta_4 = 72.2°$, and point $P = (-1.17, 2.92)$ in. relative to the origin located at bearing A_o. The construction arcs were left on the sketch to help clarify the construction procedure.

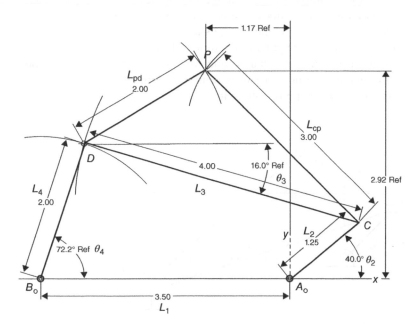

Figure 2.2   Graphical solution of 4-bar

2.2.2 *Graphical Position Analysis for a Slider-Crank Linkage*

Assume we want to determine the proper orientation for the slider-crank linkage shown in Figure 2.3 when the input, link 2, is at 115°.

Given: The origin is located at bearing A_o. Point D is 30 mm below bearing A_o. The input, link 2, is 50 mm long. Link 3 is 185 mm long. Distance from point C to point P is 75 mm. Distance from point D to point P is 145 mm. What is the angular orientation of link 3? What is the horizontal distance from the origin to point D? What is the (x, y) location of point P relative to the origin?

Procedure:

1. Locate the origin and label it A_o. Select an appropriate scale such as 1 model unit = 1 mm (actual size).
2. Since point D is below the origin and travels along a horizontal line, draw a vertical line from point A_o the length of L_1, then a horizontal line that will indicate the possible locations of point D.
3. Draw a line starting at A_o at a 115° angle above the x-axis, then mark off it length of 50 model units. Mark this point C.
4. Set your compass at the length of link 3, 185 model units. Draw an arc centered at point C so that it crosses the horizontal line which indicates possible locations for point D. The intersection is point D.
5. Draw in link 3, and then draw a rectangle centered at point D to represent the slider.
6. Set your compass at the length of L_{cp}, 75 model units. Draw an arc centered at point C.
7. Set your compass at the length of L_{pd}, 145 model units. Draw an arc centered at point D so it intersects the previously drawn arc. The intersection of these two arcs is point P. (Note there are two intersection points; choose the proper intersection point.)
8. Draw in the sides, L_{cp} and L_{pd}, to complete the slider-crank linkage.
9. Using a protractor, measure the angular orientation of link 3.
10. Using your scale, measure the horizontal distance from the origin at point A_o to point D.
11. Measure the horizontal and vertical distances from the origin to point P.
12. Box in your answers with the appropriate units.

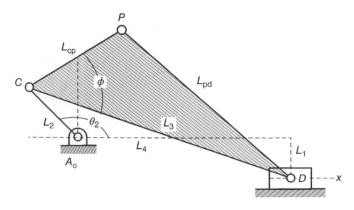

Figure 2.3 Slider-crank sketch

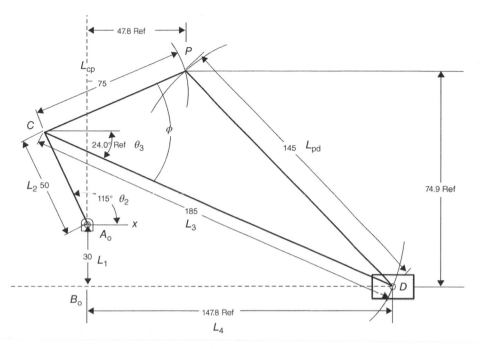

Figure 2.4   Slider-crank graphical position

From the graphical solution shown in Figure 2.4, it can be seen that $\theta_3 = -24.0°$, $L_4 = $ 147.8 mm, and point $P = (47.8$ mm, 74.9 mm) relative to the origin located at bearing A_o. The construction arcs were left on the sketch to help clarify the construction procedure.

If you wanted to determine the angular position of link 3 and the linear position of the slider relative to the origin at A_o, you would need to redraw the figure starting at step 3. For each position of link 2 that you want to analyze, you would need to redraw the figure. A better way to analyze the positions of the slider-crank linkage for varying angular positions of link 2 would be to design equations that define its position as a function of link 2's angular position.

2.3   Vector Loop Position Analysis

2.3.1   What Is a Vector?

A Euclidean vector is a geometric entity having a magnitude and a direction. In engineering, Euclidean vectors are used to represent physical quantities that have both magnitude and direction, such as force or velocity. In contrast, scalar quantities, such as mass or volume, have a magnitude but no direction.

A position vector is a vector representing the position of a point in a finite space in relation to a reference point and a coordinate system (see Figure 2.5). A displacement vector is a vector that specifies the change in position of a point relative to its previous position.

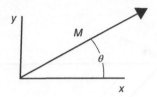

Figure 2.5   2D vector

2.3.2   Finding Vector Components of $M\angle\theta$

For a 2D vector you need its magnitude, M, and its angle relative to the positive x-axis, θ. In this textbook, all 2D vectors will be defined from the positive x-axis with the angle being positive when measured in the counterclockwise (c.c.w.) direction. A negative angle is defined as the angle from the positive x-axis measured in the clockwise (c.w.) direction.

When you are looking for (X, Y) components of a vector, the following is always true.

$$X = \text{magnitude of } x\text{-axis component vector} = M\cos\theta$$
$$Y = \text{magnitude of } y\text{-axis component vector} = M\sin\theta$$

When adding vectors, you first need to find the (X, Y) components of the vectors to be added, and then add the X-components and add the Y-components as shown below. Finally, combine the components back into a magnitude and an angle.

$$\overline{V}_{\text{sum}} = \overline{V}_1 + \overline{V}_2$$

$$\overline{V}_1 = M_1\angle\theta_1 = \{X_1, Y_1\} = \{M_1\cos\theta_1, M_1\sin\theta_1\}$$

$$\overline{V}_2 = M_2\angle\theta_2 = \{X_2, Y_2\} = \{M_2\cos\theta_2, M_2\sin\theta_2\}$$

$$X = X_1 + X_2$$

$$Y = Y_1 + Y_2$$

$$\overline{V}_{\text{sum}} = (X, Y) = M\angle\theta$$

The magnitude of the new vector is $M = \sqrt{X^2 + Y^2}$.

The angle of the new vector is $\theta = \tan^{-1}\left(\dfrac{Y}{X}\right) = a\tan 2\,(Y, X)$ {atan2 preferred}.

Care must be taken here to ensure that the angle is properly defined based on the statements above. Note that the function $\tan^{-1}(Y/X)$ returns an angle in the first or fourth quadrant only or an angle between $+90°$ and $-90°$ or $+\pi/2$ and $-\pi/2$. If the vector lies in the second or third quadrant, $\tan^{-1}(Y/X)$ does not return the correct angle. On the other hand, the function atan2(Y, X) returns an angle between $+180°$ and $-180°$ or $+\pi$ and $-\pi$, thus its answer is always in the correct quadrant.

2.3.2.1   MATLAB Vectors

In MATLAB, a 2D vector is represented by a complex number, $a + ib$. The x-component is the real part of the complex number. The y-component is the imaginary part of the complex number. In MATLAB, "i" and "j" represent the square root of minus one unless they are redefined by you as a variable in the MATLAB code. Text after a percent sign is treated as a comment and is ignored. Thus $\overline{V_{sum}} = (X,Y) = M\angle\theta$

$$\gg V1 = X1 + i*Y1$$

$$\gg V2 = X2 + i*Y2$$

$$\gg Vsum = V1 + V2 \quad \%\{\text{adding vectors}\}$$

$$\gg Vsum = X + i*Y$$

$$\gg M = abs(Vsum)$$

$$\gg theta = angle(Vsum) \quad \%\{\text{in radians}\}$$

A 2D vector can also be created as a complex number by using the complex function. The complex function has two arguments; the first is the real value and the second is the imaginary value.

$$\gg V1 = complex(X1, Y1)$$

2.3.2.2 EES Vectors

In the Engineering Equation Solver (EES), a 2D vector can also be represented by a complex number. The x-component is the real part of the complex number. The y-component is the imaginary part of the complex number. Comments ignored by the computer can be enclosed in brackets. To use complex numbers in EES, the user must enable complex number as shown below. The function $Complex allows the first argument to be "On" or "Off," and the second argument to be "i" or "j," which represents the square root of minus one.

$$\$Complex\ On\ j \qquad \{\text{complex number turned on using }"j"\}$$

$$V1 = X1 + Y1*j$$

$$V2 = X2 + Y2*j$$

$$Vsum = V1 + V2 \qquad \{\text{adding vectors}\}$$

$$Vsum = X + Y*j \qquad \{\text{answer in this form}\}$$

$$M = abs(Vsum)$$

$$thetar = anglerad(Vsum) \quad \{\text{in radians}\}$$

$$thetad = angle\deg(Vsum) \quad \{\text{in degrees}\}$$

$$\$Complex\ Off \qquad \{\text{complex numbers turned off}\}$$

A 2D vector can be represented in one of five ways in EES once $Complex On is specified as shown below. The angle can be in degrees or radians. The function cis() stands for cos + i*sin.

$$Vsum = 3 + 4*j$$

$$Vsum = 5 < 53.13\,\text{deg} \qquad \{\text{default unit for angles is degrees}\}$$

$$Vsum = 5*cis(53.13) \qquad \{cis(\phi) = \cos(\phi) + j*\sin(\phi)\}$$

$$Vsum = 5 < 0.9273\,rad$$

$$Vsum = 5*cis(0.9273\,rad) \quad \{cis(\phi) = \cos(\phi) + j*\sin(\phi)\}$$

2.3.3 Position Analysis of 4-Bar Linkage

This section describes three different techniques for determining the angular positions of links 3 and 4. The first method uses the EES software and the nonlinear X and Y component equations of the vector loop equation. The second method uses the vector handling capabilities of MATLAB to determine the angular positions of links 3 and 4. The third method uses algebra and trigonometry to create several equations that are easy to solve using your calculator.

Using Vector Loop approach for position analysis of a 4-bar linkage (Figure 2.6) leads to:

$$\overline{L_2} + \overline{L_3} - \overline{L_4} - \overline{L_1} = 0$$

or

$$\overline{L_2} + \overline{L_3} = \overline{L_1} + \overline{L_4}$$

Writing the equations for the y-component of each vector and then the x-component of each vector, we get:

$$y:\quad L_2 \sin\theta_2 + L_3 \sin\theta_3 = L_1 \sin\theta_1 + L_4 \sin\theta_4$$

$$x:\quad L_2 \cos\theta_2 + L_3 \cos\theta_3 = L_1 \cos\theta_1 + L_4 \cos\theta_4$$

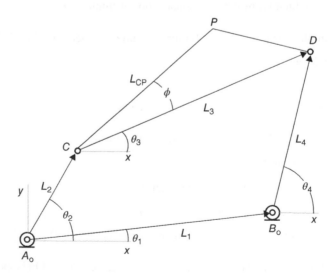

Figure 2.6 4-Bar vector loop

Based on the assumption that the location of the bearings at A_o and B_o are defined along with the size of links 2, 3, and 4, the two unknowns θ_3 and θ_4 can be determined. Since link 2 is assumed the input link, its angular position is known, θ_2. Rearranging the two equations so that the unknowns are on the left and the two known values are on the right, we get:

$$y: \quad L_3 \sin\theta_3 - L_4 \sin\theta_4 = L_1 \sin\theta_1 - L_2 \sin\theta_2$$

$$x: \quad L_3 \cos\theta_3 - L_4 \cos\theta_4 = L_1 \cos\theta_1 - L_2 \cos\theta_2$$

2.3.3.1 EES Analysis

However, if EES is used, then this step is not necessary. You are ready to solve the two equations once you determine the initial guesses for θ_3 and θ_4. These guesses come from a sketch of the linkage in its current position. If the linkage is drawn roughly to scale in Figure 2.6, then initial guesses might be $\theta_3 = 30°$ and $\theta_4 = 75°$. Anything close to these guesses should arrive at the correct answer in EES.

Example 2.1   Position Analysis of a 4-Bar Linkage Using Vector Loop Method and EES

Problem: Given a 4-bar linkage with link lengths of $L_1 - 3.00$ in., $L_2 = 1.25$ in., $L_3 = 3.25$ in., and $L_4 = 2.00$ in. Bearing B_o is located at a $10°$ angle from bearing A_o. For the current position of $\theta_2 = 60°$, determine the angles θ_3 and θ_4 as shown in Figure 2.6.

Solution: Assuming Figure 2.6 is drawn approximately to scale, the guesses for θ_3 and θ_4 can be determined as $\theta_3 = 30°$ and $\theta_4 = 75°$. Use <F9> to set these guesses in EES. Be sure trig functions are set to degrees in EES (see Figure 2.7).

Answer: $\boxed{\theta_3 = 24.3°}$ and $\boxed{\theta_4 = 71.5°}$.

If it is desired to obtain the crossed 4-bar solution, then only the initial guesses for θ_3 and θ_4 need to change. From Figure 2.8, we could guess $\theta_3 = -50°$ and $\theta_4 = -100°$.

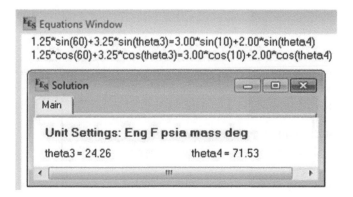

Figure 2.7   EES analysis

Example 2.2   Position Analysis of a Crossed 4-Bar Linkage Using Vector Loop Method and EES

Problem: Given a 4-bar linkage with link lengths of $L_1 = 3.00$ in., $L_2 = 1.25$ in., $L_3 = 3.25$ in., and $L_4 = 2.00$ in. Bearing B_o is located at a 10° angle from bearing A_o. For the current position of $\theta_2 = 60°$, determine the angles θ_3 and θ_4 as shown in Figure 2.8.

Solution: Assuming Figure 2.8 is drawn approximately to scale, the guesses for θ_3 and θ_4 can be determined as $\theta_3 = -50°$ and $\theta_4 = -100°$. Use <F9> to set these guesses in EES. Be sure trig functions are set to degrees in EES (see Figure 2.9).

Answer: $\boxed{\theta_3 = -51.4°}$ and $\boxed{\theta_4 = -98.6°}$.

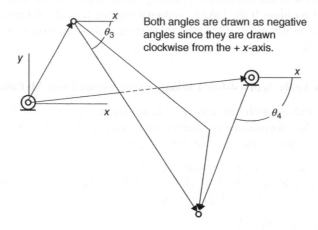

Figure 2.8   4-Bar linkage in crossed orientation

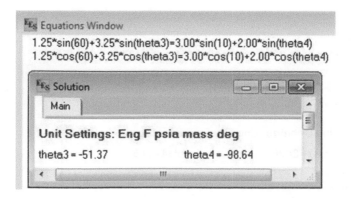

Figure 2.9   EES analysis (crossed 4-bar)

2.3.3.2 MATLAB Analysis

If MATLAB is used, then you need to take a different approach because the Student version of MATLAB cannot solve simultaneous nonlinear equations. However, you can use the function "fsolve" to solve simultaneous nonlinear equations in MATLAB if you have the Optimization Toolkit. Without "fsolve," your solution might appear as follows for the uncrossed 4-bar (see Figure 2.10). Note that $\overline{L_{bc}} = \overline{L_2} - \overline{L_1}$.

$$\gg V1 = complex(L1 * \cos d(theta1), L1 * \sin d(theta1));$$

$$\gg V2 = complex(L2 * \cos d(theta2), L2 * \sin d(theta2));$$

$$\gg Vbc = V2 - V1; \qquad\qquad \%\{V1 + Vbc = V2\}$$

$$\gg L\_bc = abs(Vbc); \qquad\qquad \%\{Length\}$$

$$\gg theta\_bc = angle(Vbc) * 180/pi(); \quad \%\{in\ degrees\}$$

As Link 2 rotates through 360°, the angle ϕ_{cd} in Figure 2.10 stays between 0° and 180° (0 and π). Since the inverse cosine returns an angle between 0 and π, it is a good choice for the solution of ϕ_{cd}. Using the cosine law, the angle ϕ_{cd} can be solved as following.

$$L_3^2 = L_{bc}^2 + L_4^2 - 2L_{bc}L_4 \cos(\phi_{cd})$$

or

$$\phi_{cd} = \cos^{-1}\left(\frac{L_{bc}^2 + L_4^2 - L_3^2}{2L_{bc}L_4}\right) \quad \{range\ =\ 0 \rightarrow +\pi\}$$

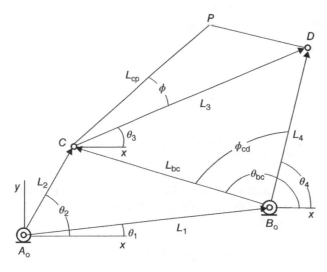

Figure 2.10 Uncrossed 4-bar vector loop with Lbc

Note that in the standard non-crossed configuration, the angle ϕ_{cd} will typically be between 0 and 180° or π radians. Completing the analysis leads to the following with the calculation for L_3 being a check for your work.

$$\theta_4 = \theta_{bc} - \phi_{cd}$$

$$\overline{L_{3\_check}} = \overline{L_4} - \overline{L_{bc}}$$

Moving back to MATLAB produces the following code where L3_check must come out to be the same as the original L3 value. If not, there is an error in your solution.

```
>> phi_cd = acosd((L_bc^2 + L4^2 - L3^2)/(2*L_bc*L4));
>> theta4 = theta_bc - phi_cd;
>> V4 = complex(L4*cosd(theta4), L4*sind(theta4));
>> V3 = V4 - Vbc;
>> theta3 = angle(V3)*180/pi();
>> L3_check = abs(V3);
```

Example 2.3   Position Analysis of a 4-Bar Linkage Using Vector Loop Method and MATLAB

Problem: Given a 4-bar linkage with link lengths of $L_1 = 3.00$ in., $L_2 = 1.25$ in., $L_3 = 3.25$ in., and $L_4 = 2.00$ in. Bearing B_o is located at a 10° angle from bearing A_o. For the current position of $\theta_2 = 60°$, determine the angles θ_3 and θ_4 as shown in Figure 2.10.

Solution: Assuming Figure 2.10 is drawn approximately to scale, the guesses for θ_3 and θ_4 can be determined as $\theta_3 = 30°$ and $\theta_4 = 75°$ (see Figure 2.11).

Answer: $\boxed{\theta_3 = 24.3°}$ and $\boxed{\theta_4 = 71.5°}$.

If it is desired to obtain the crossed 4-bar solution, then a procedure similar to the one above is used as shown below.

$$\phi_{cd} = \cos^{-1}\left(\frac{L_{bc}^2 + L_4^2 - L_3^2}{2L_{bc}L_4}\right) \quad \{\text{range} = 0 \rightarrow +\pi\}$$

Note that in the crossed configuration (Figure 2.12), the angle ϕ_{cd} will typically be between 0 and 180° or π radians. Completing the analysis leads to the following with the calculation for L_3 being a check for your work.

$$\theta_4 = \theta_{bc} + \phi_{cd} - 360$$

$$L_{3\_check} = L_4 - L_{bc}$$

MATLAB code

```
format compact
V1=complex(3.00*cosd(10.),3.00*sind(10.));
V2=complex(1.25*cosd(60.),1.25*sind(60.));
Vbc=V2-V1;
L_bc=abs(Vbc);
theta_bc=angle(Vbc)*180/pi();
phi_cd=acosd((L_bc^2+2.00^2-3.25^2)/(2*L_bc*2.00));
theta4=theta_bc-phi_cd
V4=complex(2.00*cosd(theta4),2.00*sind(theta4));
V3=V4-Vbc
theta3=angle(V3)*180/pi()
L3_check=abs(V3)
```

MATLAB solution

```
>> Fourbar_Position
theta4 =
      71.5322
theta3 =
      24.2612
L3_check =
      3.2500
>> |
```

Figure 2.11   MATLAB analysis

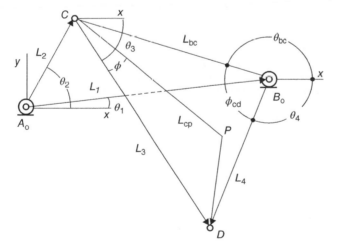

Figure 2.12   Crossed 4-bar

Moving back to MATLAB, and noting that the code is only slightly different, produces the following where L3_check must come out to be the same as the original L3 value. If not, there is an error in your solution.

$$\gg phi\_cd = a\cos d((L\_bc^\wedge 2 + L4^\wedge 2 - L3^\wedge 2)/(2*L\_bc*L4));$$

$$\gg theta4 = theta\_bc + phi\_cd - 360;$$

$$\gg V4 = complex(L4*\cos d(theta4), L4*\sin d(theta4));$$

$$\gg V3 = V4 - Vbc;$$

$$\gg theta3 = angle(V3)*180/pi();$$

$$\gg L3\_check = abs(V3);$$

Example 2.4 Position Analysis of the crossed 4-bar Linkage using Vector Loop Method and MATLAB

Problem: Given a 4-bar linkage with link lengths of $L_1 = 3.00$ in., $L_2 = 1.25$ in., $L_3 = 3.25$ in., and $L_4 = 2.00$ in. Bearing B_o is located at a $10°$ angle from bearing A_o. For the current position of $\theta_2 = 60°$, determine the angles θ_3 and θ_4 as shown in Figure 2.12.

Solution: Assuming Figure 2.12 is drawn approximately to scale, the guesses for θ_3 and θ_4 can be determined as $\theta_3 = -50°$ and $\theta_4 = -100°$ (see Figure 2.13).

Answer: $\boxed{\theta_3 = -51.4°}$ and $\boxed{\theta_4 = -98.6°}$.

```
MATLAB code
format compact
V1=complex(3.00*cosd(10.),3.00*sind(10.));
V2=complex(1.25*cosd(60.),1.25*sind(60.));
Vbc=V2-V1;
L_bc=abs(Vbc);
theta_bc=angle(Vbc)*180/pi();
phi_cd=acosd((L_bc^2+2.00^2-3.25^2)/(2*L_bc*2.00));
theta4=theta_bc+phi_cd-360
V4=complex(2.00*cosd(theta4),2.00*sind(theta4));
V3=V4-Vbc;
theta3=angle(V3)*180/pi()
L3_check=abs(V3)


MATLAB solution
>> Fourbar_Position1c
theta4 =
   -98.6412
theta3 =
   -51.3701
L3_check =
   3.2500
>> |
```

Figure 2.13 MATLAB analysis 2

If the version of MATLAB you are using contains the function "fsolve" from the Optimization Toolkit, then the following procedure can be used to solve the two unknown angles.

1. Write a function that contains the initial guesses for the unknowns as a row vector, and then call the "fsolve" function while passing the function that contains the system of nonlinear equations to solve and the initial guesses.
2. Follow this function with a new function (same m-file) that contains the given information and the vector loops equations to be solved.
3. Save the file using the name of the first function, such as "TwoEqns.m."
4. Run the MATLAB code by typing the first function name in the command window.
5. Record the answers.

Example 2.5   Position Analysis of a 4-Bar Linkage Using Vector Loop Method and "fsolve" Function

Problem: Given a 4-bar linkage with link lengths of $L_1 = 3.00$ in., $L_2 = 1.25$ in., $L_3 = 3.25$ in., and $L_4 = 2.00$ in. Bearing B_o is located at a $10°$ angle from bearing A_o. For the current position of $\theta_2 = 60°$, determine the angles θ_3 and θ_4 as shown in Figure 2.10.

Solution: Assuming Figure 2.10 is drawn approximately to scale, the guesses for θ_3 and θ_4 can be determined as $\theta_3 = 30°$ and $\theta_4 = 75°$ (see Figure 2.14).

Answers: $\boxed{\theta_3 = 19.4°}$ and $\boxed{\theta_4 = 70.7°}$.

2.3.3.3   Closed Form Analysis

If a closed form solution is desired where you can determine the position using your calculator, then the following approach might be used (see Figure 2.15).

Writing the equations for the y-component of each vector, then the x-component of each vector.

$$y: \quad L_2 \sin\theta_2 + L_3 \sin\theta_3 = L_1 \sin\theta_1 + L_4 \sin\theta_4$$
$$x: \quad L_2 \cos\theta_2 + L_3 \cos\theta_3 = L_1 \cos\theta_1 + L_4 \cos\theta_4$$

Moving the unknown angle θ_3 to the left side of the equation and everything else to the right side of the equation leads to:

$$y: \quad L_3 \sin\theta_3 = L_1 \sin\theta_1 - L_2 \sin\theta_2 + L_4 \sin\theta_4$$
$$x: \quad L_3 \cos\theta_3 = L_1 \cos\theta_1 - L_2 \cos\theta_2 + L_4 \cos\theta_4$$

Now squaring both sides of the each equation and then adding the two equations together, we get:

```
  TwoEqns.m  ⊠  +
 1      ⊟ function TwoEqns
 2 —      guess34=[20, 70];
 3 —      theta34=fsolve(@eqns,guess34)
 4
 5      ⊟ function theta=eqns(x)
 6 —      th3=x(1);
 7 —      th4=x(2);
 8 —      L1=10;
 9 —      th1=0;
10 —      L2=3;
11 —      th2=45;
12 —      L3=10;
13 —      L4=8;
14 —      theta=[L2*cos(th2)+L3*cos(th3)-L4*cos(th4)-L1*cos(th1);
15 —      L2*sin(th2)+L3*sin(th3)-L4*sin(th4)-L1*sin(th1)];
```

Command Window

```
>> TwoEqns

Equation solved.

fsolve completed because the vector of function values is near zero
as measured by the default value of the function tolerance, and
the problem appears regular as measured by the gradient.

<stopping criteria details>

theta34 =
   19.4256    70.6906
```

$$\boxed{\theta_3 = 19.4^\circ \quad \theta_4 = 70.7^\circ}$$

Figure 2.14 Solution using "fsolve" function

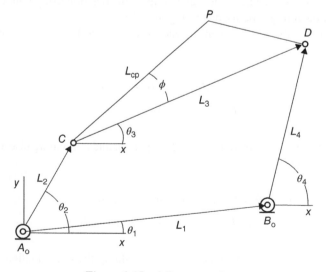

Figure 2.15 4-Bar vector loop

$$L_3^2\sin^2\theta_3 + L_3^2\cos^2\theta_3 = (L_1\sin\theta_1 - L_2\sin\theta_2 + L_4\sin\theta_4)^2 + (L_1\cos\theta_1 - L_2\cos\theta_2 + L_4\cos\theta_4)^2$$

Note that the left side of the equation can be reduced since $\sin^2\theta_3 + \cos^2\theta_3 = 1$. Then expanding the right side of the equation, we get:

$$
\begin{aligned}
L_3^2 = &(L_1^2\cos^2\theta_1 - 2L_1L_2\cos\theta_1\cos\theta_2 + 2L_1L_4\cos\theta_1\cos\theta_4 + L_2^2\cos^2\theta_1 \\
&- 2L_2L_4\cos\theta_2\cos\theta_4 + L_4^2\cos^2\theta_4) + (L_1^2\sin^2\theta_1 - 2L_1L_2\sin\theta_1\sin\theta_2 \\
&+ 2L_1L_4\sin\theta_1\sin\theta_4 + L_2^2\sin^2\theta_1 - 2L_2L_4\sin\theta_2\sin\theta_4 + L_4^2\sin^2\theta_4)
\end{aligned}
$$

Collecting like terms, we get:

$$
\begin{aligned}
L_3^2 = &L_1^2 + L_2^2 + L_4^2 - 2L_1L_2(\cos\theta_1\cos\theta_2 + \sin\theta_1\sin\theta_2) \\
&+ 2L_1L_4(\cos\theta_1\cos\theta_4 + \sin\theta_1\sin\theta_4) - 2L_2L_4(\cos\theta_2\cos\theta_4 + \sin\theta_2\sin\theta_4)
\end{aligned}
$$

Note that $\cos(\theta_1 - \theta_2) = (\cos\theta_1\cos\theta_2 + \sin\theta_1\sin\theta_2)$ and dividing each term by $2L_4$ leads to:

$$
\begin{aligned}
0 = &\frac{L_1^2 + L_2^2 + L_4^2 - L_3^2}{2L_4} - \frac{L_1L_2\cos(\theta_1 - \theta_2)}{L_4} \\
&+ L_1(\cos\theta_1\cos\theta_4 + \sin\theta_1\sin\theta_4) - L_2(\cos\theta_2\cos\theta_4 + \sin\theta_2\sin\theta_4)
\end{aligned}
$$

Collecting $\cos\theta_4$ and $\sin\theta_4$ terms, we get:

$$
\begin{aligned}
0 = &\frac{L_1^2 + L_2^2 + L_4^2 - L_3^2}{2L_4} - \frac{L_1L_2\cos(\theta_1 - \theta_2)}{L_4} \\
&+ \cos\theta_4(L_1\cos\theta_1 - L_2\cos\theta_2) + \sin\theta_4(L_1\sin\theta_1 + L_2\sin\theta_2)
\end{aligned}
$$

Now let

$$C_1 = \frac{L_1^2 + L_2^2 + L_4^2 - L_3^2}{2L_4} - \frac{L_1L_2\cos(\theta_1 - \theta_2)}{L_4}$$

$$C_2 = L_1\cos\theta_1 - L_2\cos\theta_2$$

$$C_3 = L_1\sin\theta_1 - L_2\sin\theta_2$$

Rewriting $\cos\theta_4$ and $\sin\theta_4$ in terms of the tangent function, we get:

$$\cos\theta_4 = \frac{1 - \tan^2\left(\dfrac{\theta_4}{2}\right)}{1 + \tan^2\left(\dfrac{\theta_4}{2}\right)}$$

$$\sin\theta_4 = \frac{2\tan\left(\dfrac{\theta_4}{2}\right)}{1 + \tan^2\left(\dfrac{\theta_4}{2}\right)}$$

Substituting, we get:

$$0 = C_1 + C_2 \frac{1 - \tan^2\left(\frac{\theta_4}{2}\right)}{1 + \tan^2\left(\frac{\theta_4}{2}\right)} + C_3 \frac{2\tan\left(\frac{\theta_4}{2}\right)}{1 + \tan^2\left(\frac{\theta_4}{2}\right)}$$

Multiplying through by $1 + \tan^2\left(\frac{\theta_4}{2}\right)$ and collecting like terms leads to a quadratic equation for the tangent function:

$$0 = C_1\left(1 + \tan^2\left(\frac{\theta_4}{2}\right)\right) + C_2\left(1 - \tan^2\left(\frac{\theta_4}{2}\right)\right) + C_3\left(2\tan\left(\frac{\theta_4}{2}\right)\right)$$

$$0 = (C_1 - C_2)\tan^2\left(\frac{\theta_4}{2}\right) + 2C_3\tan\left(\frac{\theta_4}{2}\right) + (C_1 + C_2)$$

Using the quadratic equation provides two roots because of "±." If the two roots are complex (has imaginary terms), then the 4-bar link has no solution for these links and this θ_2.

$$\theta_4 = 2\tan^{-1}\left(\frac{-2C_3 \pm \sqrt{4C_3^2 - 4(C_1 - C_2)(C_1 + C_2)}}{2(C_1 - C_2)}\right)$$

or   (2.1)

$$\boxed{\theta_4 = 2\tan^{-1}\left(\frac{-C_3 \pm \sqrt{(C_3^2 - C_1^2 + C_2^2)}}{(C_1 - C_2)}\right)}$$

Where

$$\boxed{\begin{aligned}
C_1 &= \frac{L_1^2 + L_2^2 + L_4^2 - L_3^2}{2L_4} - \frac{L_1 L_2 \cos(\theta_1 - \theta_2)}{L_4} \\
C_2 &= L_1\cos\theta_1 - L_2\cos\theta_2 \\
C_3 &= L_1\sin\theta_1 - L_2\sin\theta_2
\end{aligned}}$$

(2.2)

Now going back to the original equations, we can solve for θ_3 by dividing the first equation by the second equation.

$$\frac{\sin\theta_3}{\cos\theta_3} = \boxed{\tan\theta_3 = \frac{L_1\sin\theta_1 + L_4\sin\theta_4 - L_2\sin\theta_2}{L_1\cos\theta_1 + L_4\cos\theta_4 - L_2\cos\theta_2}} = \frac{\Delta y}{\Delta x}$$

(2.3)

If we use the ATAN2 function, then it is assured that the angle is located in the correct quadrant; otherwise we need to pay attention to the signs of the numerator and the denominator to correctly determine the angle measured from the +x-axis.

Example 2.6   Position Analysis of a 4-Bar Linkage Using Vector
Loop Method and Closed Form Solution

Problem: Given a 4-bar linkage with link lengths of $L_1 = 3.00$ in., $L_2 = 1.25$ in., $L_3 = 3.25$ in., and $L_4 = 2.00$ in. Bearing B_o is located at a $10°$ angle from bearing A_o. For the current position of $\theta_2 = 60°$, determine the angles θ_3 and θ_4 as shown in Figure 2.15.

Solution: Assuming Figure 2.15 is drawn approximately to scale, the guesses for θ_3 and θ_4 can be determined as $\theta_3 = 30°$ and $\theta_4 = 75°$.

Because this method calculates two answers for θ_4, we need to determine both answers and then select the one that is appropriate for our configuration of the 4-bar linkage shown in the figure.

$$C_1 = \frac{L_1^2 + L_2^2 + L_4^2 - L_3^2}{2L_4} - \frac{L_1 L_2 \cos(\theta_1 - \theta_2)}{L_4}$$

$$C_1 = \frac{(3.00^2 + 1.25^2 + 2.00^2 - 3.25^2)}{2*2.00} - \frac{(3.00 * 1.25 * \cos(10° - 60°))}{2.00} = -0.2052$$

$$C_2 = L_1 \cos\theta_1 - L_2 \cos\theta_2$$

$$C_2 = 3.00 * \cos(10°) - 1.25 * \cos(60°) = 2.329$$

$$C_3 = L_1 \sin\theta_1 - L_2 \sin\theta_2$$

$$C_3 = 3.00 * \sin(10°) - 1.25 * \sin(60°) = -0.5616$$

$$\theta_4 = 2\tan^{-1}\left(\frac{-C_3 \pm \sqrt{(C_3^2 - C_1^2 + C_2^2)}}{(C_1 - C_2)}\right)$$

$$\theta_4 = 2\tan^{-1}\left(\frac{-(-0.5616) - \sqrt{(-0.5616)^2 - (-0.2052)^2 + 2.329^2}}{((-0.2052) - 2.329)}\right) = 71.53° = \theta_4$$

$$\theta_4 = 2\tan^{-1}\left(\frac{-(-0.5616) + \sqrt{(-0.5616)^2 - (-0.2052)^2 + 2.329^2}}{((-0.2052) - 2.329)}\right) = \cancel{-98.64° = \theta_4}$$

$$\Delta y = L_1 \sin\theta_1 + L_4 \sin\theta_4 - L_2 \sin\theta_2$$

$$\Delta y = 3.00 * \sin(10°) + 2.00 * \sin(71.53°) - 1.25 * \sin(60°) = 1.335$$

$$\Delta x = L_1 \cos\theta_1 + L_4 \cos\theta_4 - L_2 \cos\theta_2$$

$$\Delta x = 3.00 * \cos(10°) + 2.00 * \cos(71.53°) - 1.25 * \cos(60°) = 2.963$$

$$\theta_3 = \tan^{-1}\left(\frac{\Delta y}{\Delta x}\right) = \tan^{-1}\left(\frac{1.335}{2.963}\right) = 24.25° = \theta_3$$

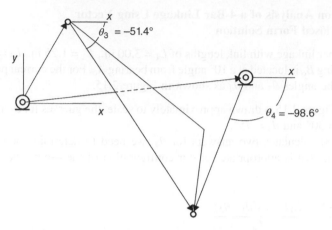

Figure 2.16   Crossed 4-bar linkage

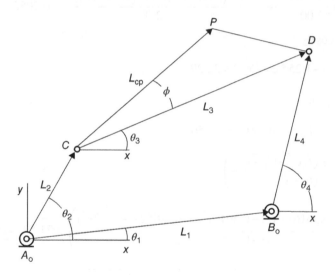

Figure 2.17   Coupler point analysis

Answer: $\boxed{\theta_3 = 24.3°}$ and $\boxed{\theta_4 = 71.5°}$.

Note that the other solution, $\theta_4 = -98.6°$ and $\theta_3 = -51.4°$, gives us the solution for the crossed 4-bar linkage (Figure 2.16).

2.3.3.4   Coupler Point Analysis

Once the angular position of links 3 and 4 has been determined, the location of the coupler point, P, is an easy extension using vectors (see Figure 2.17).

$$\overline{P} = \overline{L_2} + \overline{L_{cp}}$$

$$\overline{P} = L_2 \angle \theta_2 + L_{cp} \angle \theta_{cp}$$

or

$$\overline{P} = (P_x, P_y)$$

$$P_y = L_2 \sin \theta_2 + L_{cp} \sin \theta_{cp}$$

$$P_x = L_2 \cos \theta_2 + L_{cp} \cos \theta_{cp}$$

These same equations work for the crossed 4-bar since link 3 is defined the same way in both orientations.

Example 2.7   Coupler Point Position Analysis of a 4-Bar Linkage Using Vector Loop Method

Problem: Given a 4-bar linkage with link lengths of $L_1 = 3.00$ in., $L_2 = 1.25$ in., $L_3 = 3.25$ in., and $L_4 = 2.00$ in. Also, $L_{cp} = 2.25$ in. and $\phi = 19°$ counter clockwise. Bearing B_o is located at a 10° angle from bearing A_o. For the current position of $\theta_2 = 60°$, the angles θ_3 and θ_4 are 24.3° and 71.5° as shown in Figure 2.17. Determine the location of coupler point, P, relative to the origin at bearing A_o.

$$\overline{P} = L_2 \angle \theta_2 + L_{cp} \angle \theta_{cp}$$

$$\overline{P} = 1.25 \angle 60° + 2.25 \angle (24.26° + 19.0°)$$

$$\overline{P} = \underline{3.466 \text{ in. } \angle 49.22°}$$

or

$$\overline{P} = (P_x, P_y)$$

$$P_y = L_2 \sin \theta_2 + L_{cp} \sin \theta_{cp}$$

$$P_y = 1.25 \sin(60°) + 2.25 \sin(24.26° + 19.0°)$$

$$P_y = \underline{2.625 \text{ in.}}$$

$$P_x = L_2 \cos \theta_2 + L_{cp} \cos \theta_{cp}$$

$$P_x = 1.25 \cos(60°) + 2.25 \cos(24.26° + 19.0°)$$

$$P_x = \underline{2.264 \text{ in.}}$$

Answer: $\boxed{P = 3.47 \text{ in. } \angle 49.2°}$ or $\boxed{P = (2.26, 2.62) \text{ in.}}$

2.3.4   Position Analysis of Slider-Crank Linkage

This section describes three different techniques for determining the angular positions of link 3 and the linear position of the slider, link 4. Link 4 is in the direction of the slider motion. Link 1 is perpendicular to link 4. The first method uses the EES software and the nonlinear X and Y

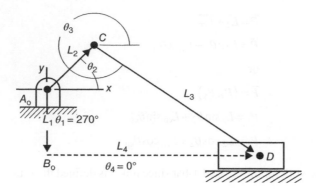

Figure 2.18   Slider-crank vector loop

component equations of the vector loop equation. The second method uses the vector handling capabilities of MATLAB to determine the angular positions of links 3 and 4. The third method uses algebra and trigonometry to create several equations that are easy to solve using your calculator.

Using Vector Loop approach for position analysis of a slider-crank linkage (Figure 2.18) leads to:

$$\overline{L_2} + \overline{L_3} - \overline{L_4} - \overline{L_1} = 0$$

or

$$\overline{L_2} + \overline{L_3} = \overline{L_1} + \overline{L_4}$$

Writing the equations for the y-component of each vector and then the x-component of each vector, we get.

$$y: \quad L_2 \sin\theta_2 + L_3 \sin\theta_3 = L_1 \sin\theta_1 + L_4 \sin\theta_4$$

$$x: \quad L_2 \cos\theta_2 + L_3 \cos\theta_3 = L_1 \cos\theta_1 + L_4 \cos\theta_4$$

Based on the assumption that the location of the bearing at A_o is defined along with the size of links 1, 2, and 3, the two unknowns θ_3 and L_4 can be determined. Since link 2 is assumed the input link, its angular position is known, θ_2. Rearranging the two equations so that the unknowns are on the left and the two known values are on the right, we get:

$$y: \quad L_3 \sin\theta_3 - L_4 \sin\theta_4 = L_1 \sin\theta_1 - L_2 \sin\theta_2$$

$$x: \quad L_3 \cos\theta_3 - L_4 \cos\theta_4 = L_1 \cos\theta_1 - L_2 \cos\theta_2$$

2.3.4.1   EES Analysis

However, if EES is used, then this step is not necessary. You are ready to solve the two equations once you determine the initial guesses for θ_3 and L_4. These guesses come from a sketch of the linkage in its current position. If the linkage is drawn roughly to scale in Figure 2.18, then initial guesses might be $\theta_3 = 330°$ and $L_4 = L_3$. Anything close to these guesses should arrive at the correct answer in EES.

Example 2.8   Position Analysis of a Slider-Crank Linkage Using Vector Loop Method and EES

Problem: Given a slider-crank linkage with link lengths of $L_1 = 2.10$ in., $L_2 = 2.00$ in., and $L_3 = 6.50$ in. Bearing A_o is located at the origin. For the current position of $\theta_2 = 43°$, determine the angle θ_3 and the length L_4 as shown in Figure 2.18. Note $\theta_1 = 270°$ and $\theta_4 = 0°$.

Solution: Assuming Figure 2.18 is drawn approximately to scale, the guesses for θ_3 and L_4 can be determined as $\theta_3 = 330°$ and $L_4 = 6.50$ in. Use <F9> to set these guesses in EES. Be sure trig functions are set to degrees in EES (see Figure 2.19).

Answer: $\boxed{\theta_3 = 328°}$ and $\boxed{L_4 = 6.96 \text{ in.}}$

If it is desired to obtain the crossed slider-crank solution, then only the initial guesses for θ_3 and L_4 need to change. From Figure 2.20, we could guess $\theta_3 = 210°$ and $L_4 = 3.00$ in.

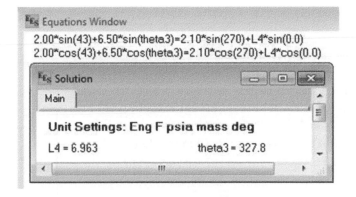

Figure 2.19   EES analysis

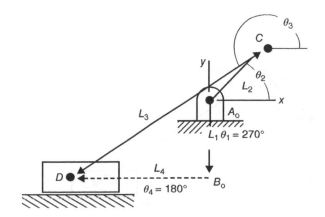

Figure 2.20   Slider-crank linkage in crossed orientation

Example 2.9   Position Analysis of a Crossed Slider-Crank Linkage Using Vector Loop Method and EES

Problem: Given a slider-crank linkage with link lengths of $L_1 = 2.10$ in., $L_2 = 2.00$ in., and $L_3 = 6.50$ in. Bearing A_o is located at the origin. For the current position of $\theta_2 = 43°$, determine the angle θ_3 and the length L_4 as shown in Figure 2.20. Note $\theta_1 = 270°$ and $\theta_4 = 180°$.

Solution: Assuming Figure 2.20 is drawn approximately to scale, the guesses for θ_3 and L_4 can be determined as $\theta_3 = 210°$ and $L_4 = 3.00$ in. Use <F9> to set these guesses in EES. Be sure trig functions are set to degrees in EES (see Figure 2.21).

Answer: $\boxed{\theta_3 = 212°}$ and $\boxed{L_4 = 4.04 \text{ in.}}$

2.3.4.2   MATLAB Analysis

If MATLAB is used and you do not have the Optimization Toolkit with "fsolve" available, then you need to take a different approach. Its solution might appear as follows for the uncrossed slider-crank linkage (see Figure 2.22). Note that $\overline{L_{bc}} = \overline{L_2} - \overline{L_1}$.

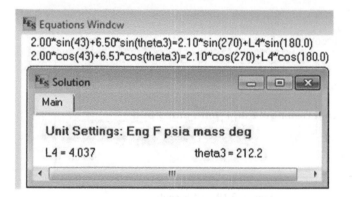

Figure 2.21   EES analysis (crossed 4-bar)

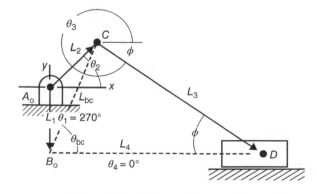

Figure 2.22   Slider-crank vector loop

$\gg V1 = complex(L1*\cos d(theta1),\quad L1*\sin d(theta1));$

$\gg V2 = complex(L2*\cos d(theta2),\quad L2*\sin d(theta2));$

$\gg Vbc = V2-V1;\qquad\qquad\qquad \%\{V1 + Vbc = V2\}$

$\gg L\_bc = abs(Vbc);\qquad\qquad\quad \%\{\text{Length from }B_o\text{ to }C\}$

$\gg theta\_bc = angle(Vbc)*180/pi();\quad \%\{\text{in degrees}\}$

Using the sine law, the angle ϕ can be solved as follows.

$$\frac{L_{bc}}{\sin\phi} = \frac{L_3}{\sin\theta_{bc}}$$

or

$$\phi = \sin^{-1}\left(\frac{L_{bc}\sin\theta_{bc}}{L_3}\right)\quad \left\{\text{range} = -\frac{\pi}{2}\rightarrow +\frac{\pi}{2}\right\}$$

Note that in the standard non-crossed configuration, the angle ϕ will typically be between $-90°$ and $+90°$ or ($-\pi/2$ and $\pi/2$ radians). Completing the analysis leads to the following with the calculation for L_4 being a check for your work.

$$\theta_3 = 360 - \phi$$

$$\overline{L_{4\_check}} = \overline{L_{bc}} + \overline{L_3}$$

Moving back to MATLAB produces the following code where L4_check must come out to have the same, original θ_4 value. If not, there is an error in your solution.

$\gg phi = a\sin d(L\_bc*\sin d(theta\_bc)/L3);$

$\gg theta3 = 360 - phi;$

$\gg V3 = complex(L3*\cos d(theta3),L3*\sin d(theta3));$

$\gg V4 = V3 + Vbc;$

$\gg L4 = abs(V4);$

$\gg theta4\_check = angle(V4)*180/pi();$

Example 2.10   Position Analysis of a Slider-Crank Linkage Using Vector Loop Method and MATLAB

Problem: Given a slider-crank linkage with link lengths of $L_1 = 2.10$ in., $L_2 = 2.00$ in., and $L_3 = 6.50$ in. Bearing A_o is located at the origin. For the current position of $\theta_2 = 43°$, determine the angle θ_3 and the length L_4 as shown in Figure 2.22. Note $\theta_1 = 270°$ and $\theta_4 = 0°$.

Solution: Assuming Figure 2.22 is drawn approximately to scale, the guesses for θ_3 and L_4 can be determined as $\theta_3 = 330°$ and $L_4 = 6.50$ in. (see Figure 2.23).

MATLAB code

```
format compact
V1=complex(2.10*cosd(270.),2.10*sind(270.));
V2=complex(2.00*cosd(43.),2.00*sind(43.));
Vbc=V2-V1;
L_bc=abs(Vbc);
theta_bc=angle(Vbc)*180/pi();
phi=asind(L_bc*sind(theta_bc)/6.50);
theta3=360-phi
V3=complex(6.50*cosd(theta3),6.50*sind(theta3));
V4=V3+Vbc;
L4=abs(V4)
theta4_check=angle(V4)*180/pi()
```

MATLAB solution

```
>> SliderCrank_Position
theta3=
    327.7969
L4=
    6.9628
theta4_check=
    3.6544e-015
>>
```

Figure 2.23   MATLAB analysis

Answer: $\boxed{\theta_3 = 328°}$ and $\boxed{L_4 = 6.96 \text{ in.}}$

If the slider is traveling at an angle other than zero degrees (Figure 2.24), we need to modify the equations slightly. Using the sine law, the angle ϕ can be solved as follows:

$$\frac{L_{bc}}{\sin\phi} = \frac{L_3}{\sin(\theta_{bc} - \theta_4)}$$

or

$$\phi = \sin^{-1}\left(\frac{L_{bc}\sin(\theta_{bc} - \theta_4)}{L_3}\right) \quad \left\{\text{range} = -\frac{\pi}{2} \rightarrow +\frac{\pi}{2}\right\}$$

Note that in the rotated configuration, the angle ϕ will typically be between −90° and +90° or (−π/2 and π/2 radians). Completing the analysis leads to the following with the calculation for L_4 being a check for your work.

$$\theta_3 = \theta_4 - \phi$$

$$\overline{L_{4\_check}} = \overline{L_{bc}} + \overline{L_3}$$

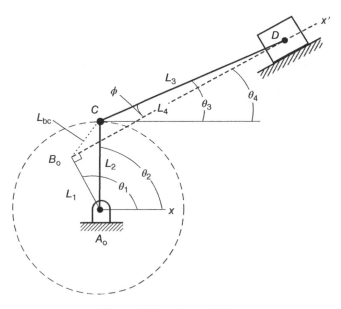

Figure 2.24 Crossed 4-bar

Moving back to MATLAB produces the following code where theta4_check must come out to have the same, original θ_4 value. If not, there is an error in your solution.

$$\gg phi = a\sin d(L\_bc * \sin d(theta\_bc - theta4)/L3);$$

$$\gg theta3 = theta4 - phi;$$

$$\gg V3 = complex(L3 * \cos d(theta3), L3 * \sin d(theta3));$$

$$\gg V4 = V3 + Vbc;$$

$$\gg L4 = abs(V4);$$

$$\gg theta4\_check = angle(V4) * 180/pi();$$

Example 2.11   Position Analysis of a Slider-Crank Linkage Using Vector Loop Method and MATLAB

Problem: Given a slider-crank linkage with link lengths of $L_1 = 40.0$ mm, $L_2 = 60.0$ mm, and $L_3 = 132$ mm. Bearing A_o is located at the origin. For the current position of $\theta_2 = 88°$, determine the angle θ_3 and the length L_4 as shown in Figure 2.24. Note $\theta_1 = 120°$ and $\theta_4 = 30°$.

Solution: Assuming Figure 2.24 is drawn approximately to scale, the guesses for θ_3 and L_4 can be determined as $\theta_3 = 30°$ and $L_4 = 150$ mm (see Figure 2.25).

Answer: $\boxed{\theta_3 = 25.3°}$ and $\boxed{L_4 = 163 \text{ mm}}$.

```
MATLAB code
format compact
V1=complex(40.0*cosd(120.),40.0*sind(120.);
V2=complex(60.0*cosd(88.),60.0*sind(120.));
Vbc=V2-V1;
L_bc=abs(Vbc);
theta_bc=angle(Vbc)*180/pi();
theta4=30.
phi=asind((L_bc*sind(theta_bc-theta4)/132.);
theta3=theta4-phi
V3=complex(132.*cosd(theta3),132.*sind(theta3));
V4=V3+Vbc;
L4=abs(V4)
theta4_check=angle(V4)*180/pi()


MATLAB solution
>> SliderCrank_Rotated
theta4=
    30
theta3=
    25.2708
L4=
    163.3458
theta4_check=
    30.0000
>>
```

Figure 2.25 MATLAB analysis 2

2.3.4.3 Closed Form Analysis

If a closed form solution for the slider-crank mechanism (Figure 2.26) is desired where you can determine the position using your calculator, then the following approach might be used.

Writing the equations for the y-component of each vector and then the x-component of each vector, we get:

$$y: \quad L_2 \sin\theta_2 + L_3 \sin\theta_3 = L_1 \sin\theta_1 + L_4 \sin\theta_4$$

$$x: \quad L_2 \cos\theta_2 + L_3 \cos\theta_3 = L_1 \cos\theta_1 + L_4 \cos\theta_4$$

Note that $\theta_1 = 270°$ and $\theta_4 = 0°$ so the above equations simplify. Also, moving the unknown angle θ_3 to the left side in the y equation and the unknown L_4 to the left in the x equation leads to:

$$y: \quad L_3 \sin\theta_3 = -L_1 - L_2 \sin\theta_2$$

$$x: \quad L_4 = L_2 \cos\theta_2 + L_3 \cos\theta_3$$

The y equation can be used to solve θ_3 and the x equation can be used to solve L_4. See equations 2.4 and 2.5.

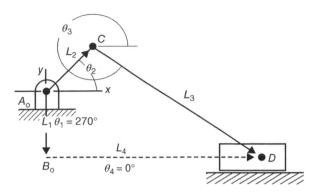

Figure 2.26 4-Bar vector loop

$$y: \quad \boxed{\theta_3 = \sin^{-1}\left(\frac{-L_1 - L_2 \sin\theta_2}{L_3}\right)} \tag{2.4}$$

$$x: \quad \boxed{L_4 = L_2 \cos\theta_2 + L_3 \cos\theta_3} \tag{2.5}$$

Note that if $\theta_1 = 90°$, then the numerator would contain $L_1 - L_2 \sin\theta_2$ and the second equation would not change.

Example 2.12 Position Analysis of a Slider-Crank Linkage Using Vector Loop Method and Closed Form Solution

Problem: Given a slider-crank linkage with link lengths of $L_1 = 2.10$ in., $L_2 = 2.00$ in., and $L_3 = 6.50$ in. Bearing A_o is located at the origin. For the current position of $\theta_2 = 43°$, determine the angle θ_3 and the length L_4 as shown in Figure 2.26. Note $\theta_1 = 270°$ and $\theta_4 = 0°$.

Solution: Assuming Figure 2.26 is drawn approximately to scale, the guesses for θ_3 and L_4 can be determined as $\theta_3 = 330°$ and $L_4 = 6.50$ in.

$$\theta_3 = \sin^{-1}\left(\frac{-L_1 - L_2 \sin\theta_2}{L_3}\right)$$

$$\theta_3 = \sin^{-1}\left(\frac{-2.10 - 2.00\sin 43°}{6.50}\right) = -32.2° \quad \text{or} \quad (327.8° = 360 - 32.2)$$

$$L_4 = L_2 \cos\theta_2 + L_3 \cos\theta_3$$

$$L_4 = 2.00\cos 43° + 6.50\cos(-32.2°) = 6.96 \,\text{inches}$$

Answer: $\boxed{\theta_3 = -32.2° \text{ or } 328°}$ and $\boxed{L_4 = 6.98 \text{ in.}}$

Note that if the slider is on an angle (Figure 2.27), the same procedure can be used if the equations are adjusted according to the rotated angle of the slider. Assume the x' axis is used

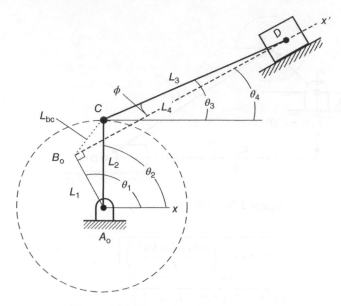

Figure 2.27   Rotated slider-crank linkage

to write the vector loop equations. Note that the numerator contains $+L_1$ since θ_1 is $(\theta_4 + 90°)$ and not $(\theta_4 - 90°)$.

$$y': \quad \theta_3 = \sin^{-1}\left(\frac{L_1 - L_2\sin(\theta_2 - \theta_4)}{L_3}\right) + \theta_4$$

$$x': \quad L_4 = L_2\cos(\theta_2 - \theta_4) + L_3\cos(\theta_3 - \theta_4)$$

Example 2.13   Position Analysis of a Slider-Crank Linkage Using Vector Loop Method and Closed Form Solution

Problem: Given a slider-crank linkage with link lengths of $L_1 = 40.0$ mm, $L_2 = 60.0$ mm, and $L_3 = 132$ mm. Bearing A_o is located at the origin. For the current position of $\theta_2 = 88°$, determine the angle θ_3 and the length L_4 as shown in Figure 2.27. Note $\theta_1 = 120°$ and $\theta_4 = 30°$.

Solution: Assuming Figure 2.27 is drawn approximately to scale, the guesses for θ_3 and L_4 can be determined as $\theta_3 = 30°$ and $L_4 = 150$ mm.

$$\theta_3 = \sin^{-1}\left(\frac{L_1 - L_2\sin(\theta_2 - \theta_4)}{L_3}\right) + \theta_4$$

$$\theta_3 = \sin^{-1}\left(\frac{40 - 60\sin(88° - 30°)}{132}\right) + 30° = 25.3°$$

$$L_4 = L_2\cos(\theta_2 - \theta_4) + L_3\cos(\theta_3 - \theta_4)$$

$$L_4 = 60\cos(88° - 30°) + 132\cos(25.3° - 30°) = 163\text{mm}$$

Answer: $\boxed{\theta_3 = 25.3°}$ and $\boxed{L_4 = 163 \text{ mm}}$.

2.3.4.4 Coupler Point Analysis

Once the angular position of link 3 and linear position of the slider, link 4, have been determined, the location of the coupler point, P, is an easy extension using vectors (see Figure 2.28).

$$\bar{P} = \overline{L_2} + \overline{L_{cp}}$$

$$\bar{P} = L_2 \angle \theta_2 + L_{cp} \angle \theta_{cp}$$

or

$$\bar{P} = (P_x, P_y)$$

$$P_y = L_2 \sin \theta_2 + L_{cp} \sin(\theta_3 + \phi)$$

$$P_x = L_2 \cos \theta_2 + L_{cp} \cos(\theta_3 + \phi)$$

These same equations work for the rotated slider-crank linkage since link 3 is defined the same way in both orientations.

Example 2.14 Coupler Point Position Analysis of a Slider-Crank Linkage Using Vector Loop Method

Problem: Given a slider-crank linkage with link lengths of $L_1 = 35.0$ mm, $L_2 = 48.0$ mm, $L_3 = 185$ mm, $\theta_1 = 270°$, and $\theta_4 = 0°$. Also, $L_{pd} = 150$ mm, $L_{cp} = 72.0$ mm, and $\phi = 50.6°$ counter clockwise. For the current position of $\theta_2 = 135°$, the angle θ_3 and the length L_4 are $-21.9°$ and 137.7 mm as shown in Figure 2.28. Determine the location of coupler point, P, relative to the origin at bearing A_o.

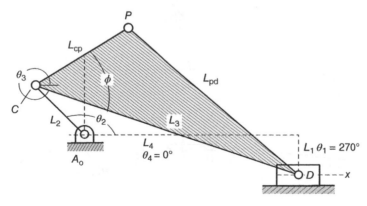

Figure 2.28   Coupler point analysis

$$\bar{P} = L_2 \angle \theta_2 + L_{cp} \angle \theta_{cp}$$

$$\bar{P} = 48.0 \angle 135° + 72.0 \angle(-21.9° + 50.6°)$$

$$\bar{P} = \underline{74.5 \, \text{mm} \angle 66.9°}$$

or

$$\bar{P} = (P_x, P_y)$$

$$P_y = L_2 \sin \theta_2 + L_{cp} \sin \theta_{cp}$$

$$P_y = 48.0 \sin(135°) + 72.0 \sin(-21.9° + 50.6°)$$

$$P_y = \underline{68.54 \, \text{mm}}$$

$$P_x = L_2 \cos \theta_2 + L_{cp} \cos \theta_{cp}$$

$$P_x = 48.0 \cos(135°) + 72.0 \cos(-21.9° + 50.6°)$$

$$P_x = \underline{29.20 \, \text{mm}}$$

Answer: $\boxed{P = 74.5 \, \text{mm} \quad \angle 66.9°}$ or $\boxed{P = (29.2, 68.5) \, \text{mm}}$.

2.3.5 Position Analysis of 6-Bar Linkage

A 6-bar linkage is typically made up of two 4-bar linkages or a 4-bar linkage and a slider-crank mechanism; therefore, it can be analyzed by analyzing the two separate linkages. No additional theory is needed to analyze a 6-bar linkage. In fact, almost any linkage can be analyzed using the procedures described for 4-bar linkages and slider-crank mechanisms.

Using Vector Loop approach for position analysis of a 6-bar linkage (Figure 2.29) leads to:

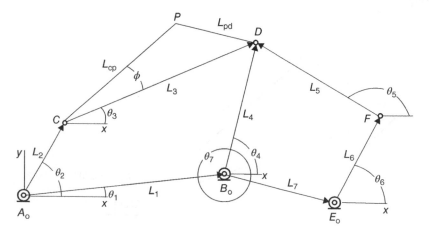

Figure 2.29 6-Bar linkage with vectors

$$\overline{L_2} + \overline{L_3} - \overline{L_4} - \overline{L_1} = 0 \quad \text{or} \quad \overline{L_2} + \overline{L_3} = \overline{L_1} + \overline{L_4}$$

and

$$\overline{L_4} - \overline{L_5} - \overline{L_6} - \overline{L_7} = 0 \quad \text{or} \quad \overline{L_4} = \overline{L_7} + \overline{L_6} + \overline{L_5}$$

Writing the equations for the y-components for both vectors and then the x-components for both vectors leads to the following.

$$y: \quad L_2 \sin\theta_2 + L_3 \sin\theta_3 = L_1 \sin\theta_1 + L_4 \sin\theta_4$$

$$y: \quad L_4 \sin\theta_4 = L_7 \sin\theta_7 + L_6 \sin\theta_6 + L_5 \sin\theta_5$$

and

$$x: \quad L_2 \cos\theta_2 + L_3 \cos\theta_3 = L_1 \cos\theta_1 + L_4 \cos\theta_4$$

$$x: \quad L_4 \cos\theta_4 = L_7 \cos\theta_7 + L_6 \cos\theta_6 + L_5 \cos\theta_5$$

If point P's location is to be determined, then the following equations can be used (see Figure 2.30).

$$\overline{P} = \overline{L_2} + \overline{L_{cp}}$$

$$P_y = L_2 \sin(\theta_2) + L_{cp} \sin(\theta_3 + \phi)$$

$$P_x = L_2 \cos(\theta_2) + L_{cp} \cos(\theta_3 + \phi)$$

Example 2.15   Position Analysis of a 6-Bar Linkage Using Vector Loop Method and EES

Problem: Given a 6-bar linkage with link lengths of $L_1 = 90.0$ mm, $L_2 = 70.0$ mm, $L_3 = 65.0$ mm, $L_4 = 75.0$ mm, $L_5 = 95.0$ mm, $L_6 = 55.0$ mm, $L_7 = 88.0$ mm, and $L_8 = 75.0$ mm. Also defined are $L_{cp} = 35.0$ mm, $L_{pd} = 49.0$ mm, and $L_{DF} = 29.0$ mm. Bearing B_o and E_o are located on the same horizontal line as bearing A_o. For the current position of $\theta_2 = 80°$, determine the angles θ_3, θ_4, θ_5, and θ_6 as shown in Figure 2.30. Also, determine the location of point P relative to the origin at bearing A_o.

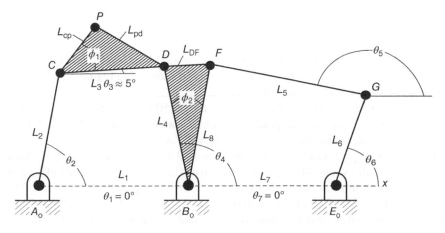

Figure 2.30   Watt 6-bar linkage

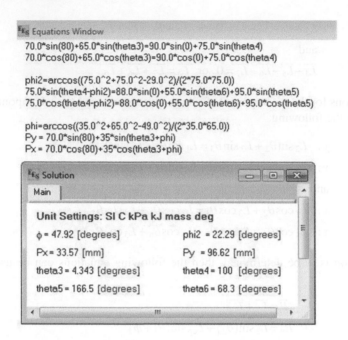

Figure 2.31   EES solution to 6-bar position analysis

Solution: Assuming Figure 2.30 is drawn approximately to scale, the guesses for θ_3 through θ_6 can be determined as $\theta_3 = 5°$, $\theta_4 = 110°$, $\theta_5 = 175°$, and $\theta_6 = 80°$. Use <F9> to set these guesses in EES. Be sure trig functions are set to degrees in EES (see Figure 2.31).

Answer: $\boxed{\theta_3 = 4.3°}$, $\boxed{\theta_4 = 100°}$, $\boxed{\theta_5 = 166.5°}$, $\boxed{\theta_6 = 68.3°}$,   and $\boxed{P = (P_x, P_y) = (33.6, 96.6) \text{ mm}}$.

Problems

2.1   Given: $L_1 = 7.25$ in., $L_2 = 2.00$ in., $L_3 = 7.00$ in., $L_4 = 4.50$ in., $\theta_1 = 5°$, and $\theta_2 = 150°$. Graphically determine the angular position of links 3 and 4 assuming Figure 2.32 is drawn roughly to scale. (Answers: $\theta_3 = 27°$, $\theta_4 = 127°$).

2.2   Given: $L_1 = 7.25$ in., $L_2 = 2.00$ in., $L_3 = 7.00$ in., $L_4 = 4.50$ in., $\theta_1 = 5°$, and $\theta_2 = 150°$. Analytically determine the angular position of links 3 and 4 assuming Figure 2.32 is drawn roughly to scale. (Answers: $\theta_3 = 27.3°$, $\theta_4 = 127.4°$).

2.3   Given: $L_1 = 7.25$ in., $L_2 = 2.00$ in., $L_3 = 7.00$ in., $L_4 = 4.50$ in., $\theta_1 = 5°$, and $\theta_2 = 150°$. What are the two lower limit extremes for the transmission angle, μ' and μ'', between links 3 and 4 if link 2 is allowed to rotate through a full 360°? (Answers: $\mu' = 48.6°$, $\mu'' = 75.0°$).

2.4   Given: $L_1 = 185$ mm, $L_2 = 50.0$ mm, $L_3 = 180$ mm, $L_4 = 115$ mm, $\theta_1 = 3°$, and $\theta_2 = 150°$. Graphically determine the angular position of links 3 and 4 assuming Figure 2.32 is drawn roughly to scale.

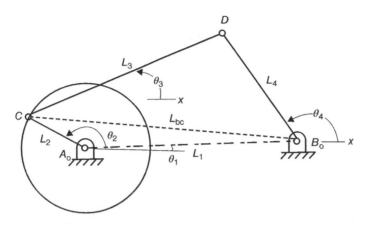

Figure 2.32 Problems 2.1 through 2.8

2.5   Given: $L_1 = 185$ mm, $L_2 = 50.0$ mm, $L_3 = 180$ mm, $L_4 = 115$ mm, $\theta_1 = 3°$, and $\theta_2 = 150°$. Analytically determine the angular position of links 3 and 4 assuming Figure 2.32 is drawn roughly to scale.

2.6   Given: $L_1 = 185$ mm, $L_2 = 50.0$ mm, $L_3 = 180$ mm, $L_4 = 115$ mm, $\theta_1 = 3°$, and $\theta_2 = 150°$. What are the two lower limit extremes for the transmission angle, μ' and μ'', between links 3 and 4 if link 2 is allowed to rotate through a full $360°$?

2.7   Given: $L_1 = 7.25$ in., $L_2 = 2.00$ in., $L_3 = 7.00$ in., $L_4 = 4.50$ in., $\theta_1 = 0°$, and $\theta_2 = 145°$. Find the angular position of links 3 and 4 assuming Figure 2.32 is drawn roughly to scale.

2.8   Given: $L_1 = 185$ mm, $L_2 = 50.0$ mm, $L_3 = 180$ mm, $L_4 = 115$ mm, $\theta_1 = 0°$, and $\theta_2 = 147°$. Find the angular position of links 3 and 4 assuming Figure 2.32 is drawn roughly to scale.

2.9   Given: $L_1 = 8.50$ in., $L_2 = 4.50$ in., $L_3 = 5.00$ in., $L_4 = 3.00$ in., $\theta_1 = 0°$, and $\theta_2 = 60°$. Find the angular position of links 3 and 4 assuming Figure 2.33 is drawn roughly to scale.

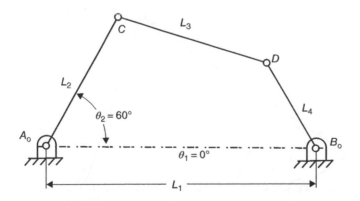

Figure 2.33 Problems 2.9 and 2.10

2.10   Given: $L_1 = 215$ mm, $L_2 = 115$ mm, $L_3 = 124$ mm, $L_4 = 75.0$ mm, $\theta_1 = 0°$, and $\theta_2 = 60°$. Find the angular position of links 3 and 4 assuming Figure 2.33 is drawn roughly to scale.

2.11   Given: $L_1 = 8.50$ in., $L_2 = 3.00$ in., $L_3 = 5.00$ in., $L_4 = 4.50$ in., $\theta_1 = 0°$, and $\theta_2 = 60°$. Graphically determine the angular position of links 3 and 4 assuming Figure 2.34 is drawn roughly to scale.

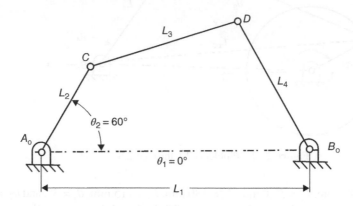

Figure 2.34   Problems 2.11 through 2.13

2.12   Given: $L_1 = 8.50$ in., $L_2 = 3.00$ in., $L_3 = 5.00$ in., $L_4 = 4.50$ in., $\theta_1 = 0°$, and $\theta_2 = 60°$. Analytically determine the angular position of links 3 and 4 assuming Figure 2.34 is drawn roughly to scale.

2.13   Given: $L_1 = 215$ mm, $L_2 = 75.0$ mm, $L_3 = 124$ mm, $L_4 = 115$ mm, $\theta_1 = 0°$, and $\theta_2 = 60°$. Find the angular position of links 3 and 4 assuming Figure 2.34 is drawn roughly to scale.

2.14   Given: $L_1 = 5.50$ in., $L_2 = 1.50$ in., $L_3 = 6.50$ in., $L_4 = 4.50$ in., $\theta_1 = 0°$, and $\theta_2 = 142°$. Find the angular position of links 3 and 4 assuming Figure 2.35 is drawn roughly to scale.

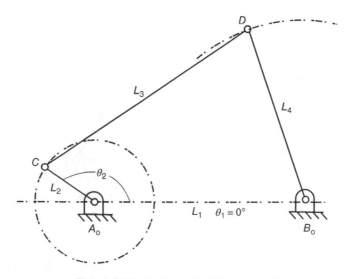

Figure 2.35   Problems 2.14 through 2.18

2.15   Given: $L_1 = 5.50$ in., $L_2 = 1.50$ in., $L_3 = 6.50$ in., $L_4 = 4.50$ in., $\theta_1 = 0°$, and $\theta_2 = 142°$. What are the two lower limit extremes for the transmission angle, μ' and μ'', between links 3 and 4 if link 2 is allowed to rotate through a full 360°?

2.16   Given: $L_1 = 140$ mm, $L_2 = 40.0$ mm, $L_3 = 160$ mm, $L_4 = 120$ mm, $\theta_1 = 0°$, and $\theta_2 = 140°$. Graphically determine the angular position of links 3 and 4 assuming Figure 2.35 is drawn roughly to scale.

2.17   Given: $L_1 = 140$ mm, $L_2 = 40.0$ mm, $L_3 = 160$ mm, $L_4 = 120$ mm, $\theta_1 = 0°$, and $\theta_2 = 140°$. Analytically determine the angular position of links 3 and 4 assuming Figure 2.35 is drawn roughly to scale.

2.18   Given: $L_1 = 140$ mm, $L_2 = 40.0$ mm, $L_3 = 160$ mm, $L_4 = 120$ mm, $\theta_1 = 0°$, and $\theta_2 = 140°$. What are the two lower limit extremes for the transmission angle, μ' and μ'', between links 3 and 4 if link 2 is allowed to rotate through a full 360°?

2.19   Given: $L_1 = 3.00$ in., $L_2 = 6.38$ in., $L_3 = 4.00$ in., $L_4 = 4.50$ in., $\theta_1 = 0°$, and $\theta_2 = 25°$. Graphically determine the angular position of links 3 and 4 assuming Figure 2.36 is drawn roughly to scale. (Answers: $\theta_3 = 155°$, $\theta_4 = 100°$).

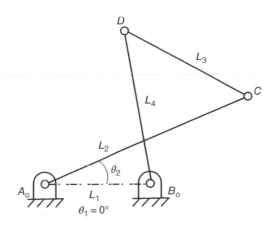

Figure 2.36   Problems 2.19 through 2.21

2.20   Given: $L_1 = 3.00$ in., $L_2 = 6.38$ in., $L_3 = 4.00$ in., $L_4 = 4.50$ in., $\theta_1 = 0°$, and $\theta_2 = 25°$. Analytically determine the angular position of links 3 and 4 assuming Figure 2.36 is drawn roughly to scale. (Answers: $\theta_3 = 154.4°$, $\theta_4 = 100.6°$).

2.21   Given: $L_1 = 80.0$ mm, $L_2 = 160$ mm, $L_3 = 100$ mm, $L_4 = 110$ mm, $\theta_1 = 0°$, and $\theta_2 = 25°$. Find the angular position of links 3 and 4 assuming Figure 2.36 is drawn roughly to scale.

2.22   Given: $L_1 = 7.38$ in., $L_2 = 3.60$ in., $L_3 = 7.25$ in., $L_4 = 3.75$ in., $\theta_1 = 0°$, and $\theta_2 = 30°$. Find the angular position of links 3 and 4 assuming Figure 2.37 is drawn roughly to scale. What are the two extremes for the transmission angle, μ' and μ'', between links 3 and 4 if link 2 is allowed to rotate through a full 360°?

2.23   Given: $L_1 = 190$ mm, $L_2 = 90.0$ mm, $L_3 = 185$ mm, $L_4 = 96.0$ mm, $\theta_1 = 0°$, and $\theta_2 = 30°$. Graphically determine the angular position of links 3 and 4 assuming Figure 2.37 is drawn roughly to scale.

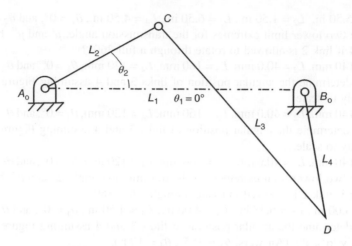

Figure 2.37   Problems 2.22 through 2.24

2.24   Given: $L_1 = 190$ mm, $L_2 = 90.0$ mm, $L_3 = 185$ mm, $L_4 = 96.0$ mm, $\theta_1 = 0°$, and $\theta_2 = 30°$. Analytically determine the angular position of links 3 and 4 assuming Figure 2.37 is drawn roughly to scale.

2.25   Given: $L_1 = 7.90$ in., $L_2 = 2.75$ in., $L_3 = 6.25$ in., $L_4 = 4.80$ in., $\theta_1 = 20°$, and $\theta_2 = 310°$. Find the angular position of links 3 and 4 assuming Figure 2.38 is drawn roughly to scale. What are the two extremes for the transmission angle, μ' and μ'', between links 3 and 4 if link 2 is allowed to rotate through a full 360°? Also, determine the (x, y) location of point D assuming the origin is located at A_o.

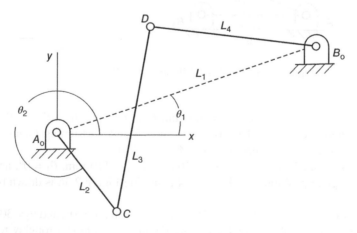

Figure 2.38   Problems 2.25 and 2.26

2.26   Given: $L_1 = 175$ mm, $L_2 = 65.0$ mm, $L_3 = 170$ mm, $L_4 = 105$ mm, $\theta_1 = 20°$, and $\theta_2 = 300°$. Find the angular position of links 3 and 4 assuming Figure 2.38 is drawn roughly to scale. What are the two extremes for the transmission angle, μ' and μ'', between links 3 and 4 if link 2 is allowed to rotate through a full 360°? Also, determine the (x, y) location of point D assuming the origin is located at A_o.

2.27   Given: $L_1 = 3.52$ in., $L_2 = 1.25$ in., $L_3 = 4.00$ in., $L_4 = 2.00$ in., $\theta_1 = 0°$, $\theta_2 = 118°$, $L_{cp} = 3.00$ in., and $L_{pd} = 2.00$ in. Find the angular position of links 3 and 4 assuming Figure 2.39 is drawn roughly to scale. Also, determine the (x, y) location of point P assuming the origin is located at A_o. (Answers: $\theta_3 = 12.8°$, $\theta_4 = 95.9°$, $P_x = 1.65$ in., $P_y = 3.10$ in.).

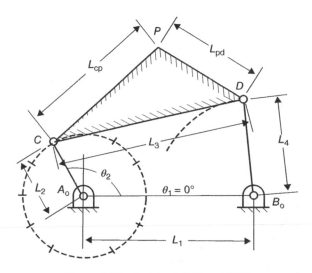

Figure 2.39   Problems 2.27 and 2.28

2.28   Given: $L_1 = 115$ mm, $L_2 = 40.0$ mm, $L_3 = 100$ mm, $L_4 = 65.0$ mm, $\theta_1 = 0°$, and $\theta_2 = 118°$, $L_{cp} = 75.0$ mm, and $L_{pd} = 50.0$ mm. Find the angular position of links 3 and 4 assuming Figure 2.39 is drawn roughly to scale. Also, determine the (x, y) location of point P assuming the origin is located at A_o. (Answers: $\theta_3 = 11.0°$, $\theta_4 = 123.2°$, $P_x = 38.7$ mm, $P_y = 83.5$ mm).

2.29   Given: $L_1 = 3.52$ in., $L_2 = 1.25$ in., $L_3 = 4.00$ in., $L_4 = 2.00$ in., $L_{cp} = 0.92$ in., $\theta_1 = 0°$, and $\theta_2 = 118°$. Find the angular position of links 3 and 4 assuming Figure 2.40 is drawn roughly to scale. Also, determine the (x, y) location of point P assuming the origin is located at A_o.

2.30   Given: $L_1 = 115$ mm, $L_2 = 40.0$ mm, $L_3 = 90.0$ mm, $L_4 = 65.0$ mm, $L_{cp} = 45.0$ mm, $\theta_1 = 0°$, and $\theta_2 = 119°$. Find the angular position of links 3 and 4 assuming Figure 2.40 is drawn roughly to scale. Also, determine the (x, y) location of point P assuming the origin is located at A_o. (Answers: $\theta_3 = -36.7°$ or $323.3°$, $\theta_4 = 196.8°$, $P_x = 16.7$ mm, $P_y = 8.1$ mm).

2.31   Given: $L_1 = 7.25$ in., $L_2 = 2.75$ in., $L_3 = 6.75$ in., $L_4 = 3.10$ in., $\theta_1 = 340°$, and $\theta_2 = 98°$, $L_{cp} = 4.75$ in., and $L_{pd} = 2.88$ in. Graphically determine the angular position of links 3 and 4 assuming Figure 2.41 is drawn roughly to scale. Also, determine the (x, y) location of point P assuming the origin is located at A_o.

2.32   Given: $L_1 = 7.25$ in., $L_2 = 2.75$ in., $L_3 = 6.75$ in., $L_4 = 3.10$ in., $\theta_1 = 340°$, and $\theta_2 = 98°$, $L_{cp} = 4.75$ in., and $L_{pd} = 2.88$ in. Analytically determine the angular position of links 3 and 4 assuming Figure 2.41 is drawn roughly to scale. Also, determine the (x, y) location of point P assuming the origin is located at A_o.

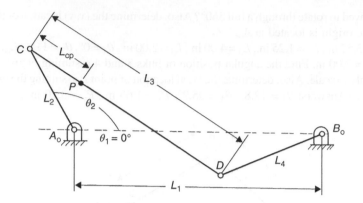

Figure 2.40   Problems 2.29 and 2.30

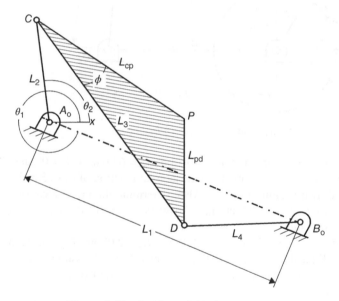

Figure 2.41   Problems 2.31 through 2.33

2.33   Given: $L_1 = 185$ mm, $L_2 = 70.0$ mm, $L_3 = 170$ mm, $L_4 = 80.0$ mm, $\theta_1 = 340°$, and $\theta_2 = 98°$, $L_{cp} = 120$ mm, and $L_{pd} = 70.0$ mm. Find the angular position of links 3 and 4 assuming Figure 2.41 is drawn roughly to scale. Also, determine the (x, y) location of point P assuming the origin is located at A_o.

2.34   Given: $L_1 = 3.50$ in., $L_2 = 3.25$ in., $L_3 = 6.50$ in., $\theta_1 = 90°$, $\theta_4 = 0°$, and currently $\theta_2 = 110°$. Graphically determine the angular position of link 3 and the linear position of link 4 (point D) relative to the origin at A_o assuming Figure 2.42 is drawn roughly to scale. (Answers: $\theta_3 = 4°$, $L_4 = 5^3/_8$ in.).

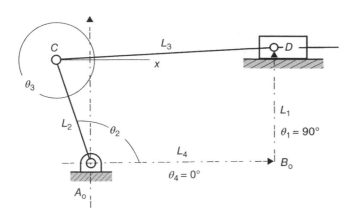

Figure 2.42   Problems 2.34 through 2.37

2.35   Given: $L_1 = 3.50$ in., $L_2 = 3.25$ in., $L_3 = 6.50$ in., $\theta_1 = 90°$, $\theta_4 = 0°$, and currently $\theta_2 = 110°$. Analytically determine the angular position of link 3 and the linear position of link 4 (point D) relative to the origin at A_o assuming Figure 2.42 is drawn roughly to scale. (Answers: $\theta_3 = 3.9°$, $L_4 = 5.37$ in.).

2.36   Given: $L_1 = 90.0$ mm, $L_2 = 84.0$ mm, $L_3 = 170$ mm, $\theta_1 = 90°$, $\theta_4 = 0°$, and currently $\theta_2 = 110°$. Graphically determine the angular position of link 3 and the linear position of link 4 (point D) relative to the origin at A_o assuming Figure 2.42 is drawn roughly to scale.

2.37   Given: $L_1 = 90.0$ mm, $L_2 = 84.0$ mm, $L_3 = 170$ mm, $\theta_1 = 90°$, $\theta_4 = 0°$, and currently $\theta_2 = 110°$. Analytically determine the angular position of link 3 and the linear position of link 4 (point D) relative to the origin at A_o assuming Figure 2.42 is drawn roughly to scale.

2.38   Given: $L_1 = 0.75$ in., $L_2 = 2.12$ in., $L_3 = 8.00$ in., $\theta_1 = 270°$, $\theta_4 = 0°$, and currently $\theta_2 = 65°$. Find the angular position of link 3 and the linear position of link 4 (point D) relative to the origin at A_o assuming Figure 2.43 is drawn roughly to scale.

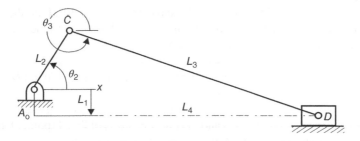

Figure 2.43   Problems 2.38 and 2.39

2.39   Given: $L_1 = 20.0$ mm, $L_2 = 50.0$ mm, $L_3 = 200$ mm, $\theta_1 = 270°$, $\theta_4 = 0°$, and currently $\theta_2 = 62°$. Find the angular position of link 3 and the linear position of link 4 (point D) relative to the origin at A_o assuming Figure 2.43 is drawn roughly to scale.

2.40   Given: $L_1 = 3.75$ in., $L_2 = 2.50$ in., $L_3 = 3.25$ in., $\theta_1 = 270°$, $\theta_4 = 0°$, and currently $\theta_2 = 240°$. Graphically determine the angular position of link 3 and the linear position of link 4 (point D) relative to the origin at A_o assuming Figure 2.44 is drawn roughly to scale.

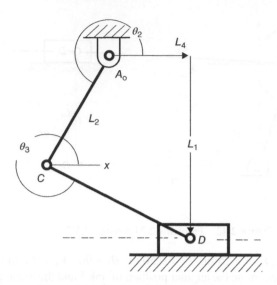

Figure 2.44 Problems 2.40 through 2.42

2.41 Given: $L_1 = 3.75$ in., $L_2 = 2.50$ in., $L_3 = 3.25$ in., $\theta_1 = 270°$, $\theta_4 = 0°$, and currently $\theta_2 = 240°$. Analytically determine the angular position of link 3 and the linear position of link 4 (point D) relative to the origin at A_o assuming Figure 2.44 is drawn roughly to scale.

2.42 Given: $L_1 = 95.0$ mm, $L_2 = 65.0$ mm, $L_3 = 85.0$ mm, $\theta_1 = 270°$, $\theta_4 = 0°$, and currently $\theta_2 = 240°$. Find the angular position of link 3 and the linear position of link 4 (point D) relative to the origin at A_o assuming Figure 2.44 is drawn roughly to scale. (Answers: $\theta_3 = 332.9°$ or $-27.1°$, $L_4 = 43.2$ mm).

2.43 Given: $L_2 = 2.25$ in., $L_3 = 5.25$ in., $\theta_4 = 90°$, and currently $\theta_2 = 32°$. Find the angular position of link 3 and the linear position of link 4 (point D) relative to the origin at A_o assuming Figure 2.45 is drawn roughly to scale.

2.44 Given: $L_2 = 60.0$ mm, $L_3 = 135$ mm, $\theta_4 = 90°$, and currently $\theta_2 = 30°$. Find the angular position of link 3 and the linear position of link 4 (point D) relative to the origin at A_o assuming Figure 2.45 is drawn roughly to scale.

2.45 Given: $L_1 = 2.75$ in., $L_2 = 2.00$ in., $L_3 = 6.50$ in., $\theta_1 = 270°$, $\theta_4 = 0°$, and currently $\theta_2 = 28°$. Find the angular position of link 3 and the linear position of link 4 (point D) relative to the origin at A_o assuming Figure 2.46 is drawn roughly to scale.

2.46 Given: $L_1 = 70.0$ mm, $L_2 = 55.0$ mm, $L_3 = 165$ mm, $\theta_1 = 270°$, $\theta_4 = 0°$, and currently $\theta_2 = 25°$. Find the angular position of link 3 and the linear position of link 4 (point D) relative to the origin at A_o assuming Figure 2.46 is drawn roughly to scale.

2.47 Given: $L_1 = 1.50$ in., $L_2 = 2.25$ in., $L_3 = 5.25$ in., $\theta_1 = 118$, $\theta_4 = 28°$, and currently $\theta_2 = 90°$. Find the angular position of link 3 and the linear position of link 4 (point D) relative to the origin at A_o assuming Figure 2.47 is drawn roughly to scale.

2.48 Given: $L_1 = 40.0$ mm, $L_2 = 62.0$ mm, $L_3 = 140$ mm, $\theta_1 = 118°$, $\theta_4 = 28°$, and currently $\theta_2 = 88°$. Find the angular position of link 3 and the linear position of link 4 (point D) relative to the origin at A_o assuming Figure 2.47 is drawn roughly to scale.

2.49 Given: $L_1 = 2.50$ in., $L_2 = 4.25$ in., $L_3 = 16.8$ in., $\theta_1 = 270°$, $\theta_4 = 0°$, $L_{cp} = 6.50$ in., $L_{pd} = 13.5$ in., and currently $\theta_2 = 135°$. Find the angular position of link 3 and the linear position of link 4 (point D) relative to the origin at A_o assuming Figure 2.48 is drawn roughly to scale. What is the angle between L_{cp} and L_3? Also, find the current location of point $P = (P_x, P_y)$ assuming A_o is the origin.

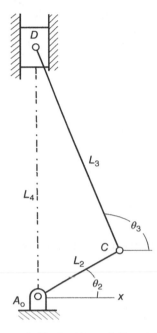

Figure 2.45   Problems 2.43 and 2.44

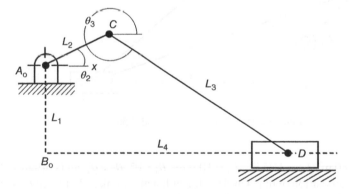

Figure 2.46   Problems 2.45 and 2.46

2.50   Given: $L_1 = 30.0$ mm, $L_2 = 55.0$ mm, $L_3 = 210$ mm, $\theta_1 = 270°$, $\theta_4 = 0°$, $L_{cp} = 84.0$ mm, $L_{pd} = 170$ mm, and currently $\theta_2 = 135°$. Find the angular position of link 3 and the linear position of link 4 (point D) relative to the origin at A_o assuming Figure 2.48 is drawn roughly to scale. What is the angle between L_{cp} and L_3? Also, find the current location of point $P = (P_x, P_y)$ assuming A_o is the origin.

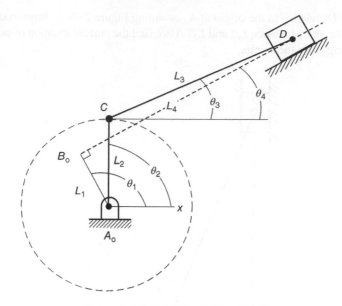

Figure 2.47   Problems 2.47 and 2.48

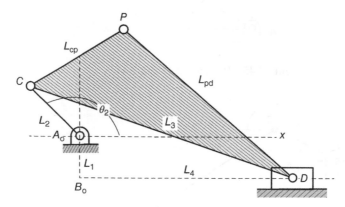

Figure 2.48   Problems 2.49 and 2.50

2.51   Given: $L_1 = 5.60$ in., $L_2 = 3.25$ in., $L_4 = 10.0$ in., $\theta_1 = 0°$, $\theta_3 = \theta_4$, and currently $\theta_2 = 66°$. Find the angular position of link 4 and the linear length of link 3 (B_o to point C) assuming Figure 2.49 is drawn roughly to scale.

2.52   Given: $L_1 = 140$ mm, $L_2 = 85.0$ mm, $L_4 = 250$ mm, $\theta_1 = 0°$, $\theta_3 = \theta_4$, and currently $\theta_2 = 66°$. Find the angular position of link 4 and the linear length of link 3 (B_o to point C) assuming Figure 2.49 is drawn roughly to scale. (Answers: $\theta_4 = 24.0°$, $L_3 = 191$ mm).

2.53   Given: $L_1 = 4.00$ in., $L_2 = 2.50$ in., $L_4 = 10.0$ in., $L_5 = 3.75$ in., $\theta_1 = \theta_5 = 0°$, $\theta_3 = \theta_4$, and currently $\theta_2 = 62°$. Find the angular position of link 4, the linear length of L3 (B_o to point C), and the vertical height of point D above the x-axis ($B_o - A_o$) assuming Figure 2.50 is drawn roughly to scale.

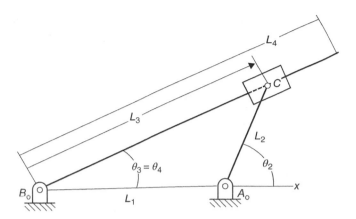

Figure 2.49   Problems 2.51 and 2.52

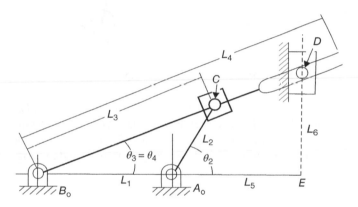

Figure 2.50   Problems 2.53 and 2.54

2.54   Given: $L_1 = 105$ mm, $L_2 = 60.0$ mm, $L_4 = 250$ mm, $L_5 = 96.0$ mm, $\theta_1 = \theta_5 = 0°$, $\theta_3 = \theta_4$, and currently $\theta_2 = 62°$. Find the angular position of link 4, the linear length of L3 (B_o to point C), and the vertical height of point D above the x-axis ($B_o - A_o$) assuming Figure 2.50 is drawn roughly to scale.

2.55   Given: $L_1 = 2.88$ in., $L_2 = 2.75$ in., $L_3 = 4.00$ in., $L_4 = 3.25$ in., $L_5 = 4.75$ in., $L_6 = 4.00$ in., $L_7 = 4.38$ in., $\theta_1 = \theta_7 = 0°$, and currently $\theta_2 = 101°$. Also, $L_{cp} = 2.50$ in., $L_{pd} = 2.37$ in., $L_{P1P2} = 3.25$ in., and $L_{P2F} = 2.38$ in. Label the angles θ_3 and θ_5 on your sketch. Find the angular positions of links 3, 4, 5, and 6. Find the current location of point $P_1 = (P_{x1}, P_{y1})$ and point $P_2 = (P_{x2}, P_{y2})$ assuming Figure 2.51 is drawn roughly to scale. Origin at A_o. (Answers: $\theta_3 = 7.2°$, $\theta_4 = 80.0°$, $\theta_5 = -6.2°$, $\theta_6 = 107.1°$, $P_{1x} = 1.36$ in., $P_{1y} = 4.34$ in., $P_{2x} = 4.40$ in., $P_{2y} = 5.50$ in.).

2.56   Given: $L_1 = 75.0$ mm, $L_2 = 70.0$ mm, $L_3 = 100$ mm, $L_4 = 80.0$ mm, $L_5 = 120$ mm, $L_6 = 100$ mm, $L_7 = 115$ mm, $\theta_1 = \theta_7 = 0°$, and currently $\theta_2 = 95°$. Also, $L_{cp} = 65.0$ mm, $L_{pd} = 60.0$ mm, $L_{P1P2} = 85.0$ mm, and $L_{P2F} = 60.0$ mm. Label the angles θ_3 and θ_5 on your sketch. Find the angular positions of links 3, 4, 5, and 6. Find the current location of point $P_1 = (P_{x1}, P_{y1})$ and point $P_2 = (P_{x2}, P_{y2})$ assuming Figure 2.51 is drawn roughly to scale. Origin at A_o.

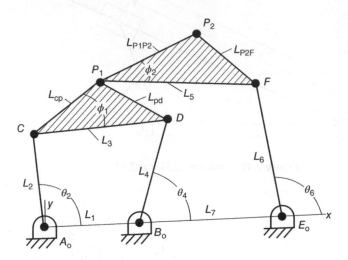

Figure 2.51   Problems 2.55 and 2.56

2.57   Given: $L_1 = 3.00$ in., $L_2 = 1.55$ in., $L_3 = 2.50$ in., $L_4 = 2.75$ in., $L_5 = 2.88$ in., $L_6 = 3.88$ in., $L_7 = 3.25$ in., $\theta_1 = 180°$, $\theta_7 = -44°$ and currently $\theta_2 = 103°$. Also, $L_{cp} = 2.88$ in. and $L_{pd} = 4.25$ in. Find the angular positions of links 3, 4, 5, and 6. Find the current location of point $P = (P_x, P_y)$ assuming Figure 2.52 is drawn roughly to scale and the origin is at B_o.

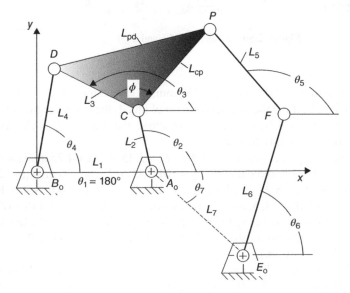

Figure 2.52   Problems 2.57 and 2.58

2.58   Given: $L_1 = 75.0$ mm, $L_2 = 40.0$ mm, $L_3 = 60.0$ mm, $L_4 = 70.0$ mm, $L_5 = 72.0$ mm, $L_6 = 95.0$ mm, $L_7 = 80.0$ mm, $\theta_1 = 180°$, $\theta_7 = -43°$ and currently $\theta_2 = 102°$. Also, $L_{cp} = 60.0$ mm and $L_{pd} = 90.0$ mm. Find the angular positions of links 3, 4, 5, and 6. Find the current location of point $P = (P_x, P_y)$ assuming Figure 2.52 is drawn roughly to scale and the origin is at B_o.

2.59   Determine the horizontal distance that the slider, link 6, is away from the origin located at B_o. $L_1 = 8.50$ in. and vertical, $L_2 = 5.75$ in. currently at 20°, $L_3 = 19.5$ in., $L_5 = 7.50$ in., and $L_7 = 14.0$ in. and vertical. What is the distance from B_o to pin C, also referred to as L_4? See Figure 2.53.

2.60   Determine the horizontal distance that the slider, link 6, is away from the origin located at B_o. $L_1 = 90$ mm and vertical, $L_2 = 60$ mm currently at 18°, $L_3 = 200$ mm, $L_5 = 80$ mm, and $L_7 = 150$ mm and vertical. What is the distance from B_o to pin C, also referred to as L_4? See Figure 2.53.

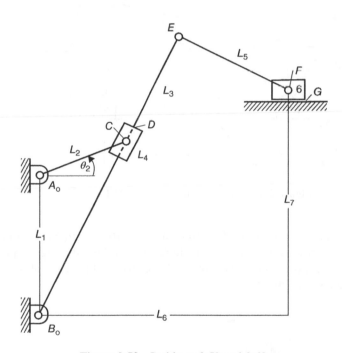

Figure 2.53   Problems 2.59 and 2.60

Programming Exercises

P2.1   Given: $L_1 = 3.52$ in., $L_2 = 1.25$ in., $L_3 = 4.00$ in., $L_4 = 2.00$ in., $\theta_1 = 0°$, $L_{cp} = 3.00$ in., and $L_{pd} = 2.00$ in. Start with $\theta_2 = 0°$ and increment it by $5°$ for a complete revolution. Find the angular position of links 3 and 4 assuming Figure 2.54 is drawn roughly to scale. Also, determine the (x, y) location of point P assuming the origin is located at A_o. Plot P_y versus P_x to show the true shape of the coupler curve generated by point P.

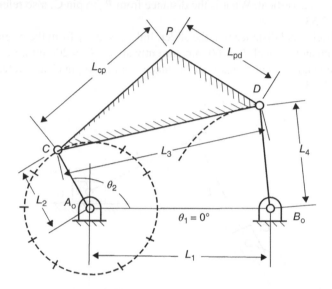

Figure 2.54   Programming Exercise P2.1

P2.2   Given: $L_1 = 70.0$ mm, $L_2 = 55.0$ mm, $L_3 = 165$ mm, $\theta_1 = 270°$, and $\theta_4 = 0°$. Start with $\theta_2 = -18.5°$ and increment it by $10°$ for a complete revolution. Find the angular position of link 3 and the linear position of link 4 (point D) relative to the origin at A_o assuming Figure 2.55 is drawn roughly to scale.

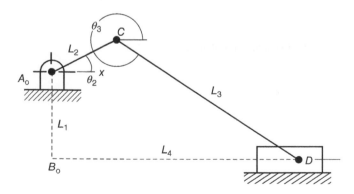

Figure 2.55   Programming Exercise P2.2

P2.3   Given: $L_1 = 30.0$ mm, $L_2 = 55.0$ mm, $L_3 = 210$ mm, $\theta_1 = 270°$, $\theta_4 = 0°$, $L_{cp} = 84.0$ mm, $L_{pd} = 170$ mm. Start with $\theta_2 = 0°$ and increment it by 5° for a complete revolution. Find the angular position of link 3 and the linear position of link 4 (point D) relative to the origin at A_o assuming Figure 2.56 is drawn roughly to scale. Also, find the current location of point $P = (P_x, P_y)$. Plot P_y versus P_x to show the true shape of the coupler curve generated by point P. Origin at A_o.

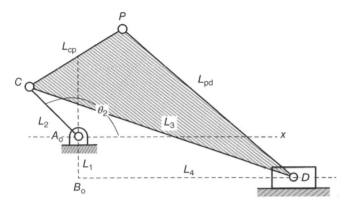

Figure 2.56   Programming Exercise P2.3

P2.4   Given: $L_1 = 4.00$ in., $L_2 = 2.50$ in., $L_4 = 10.0$ in., $L_5 = 3.75$ in., $\theta_1 = \theta_5 = 0°$, and $\theta_3 = \theta_4$. Start with $\theta_2 = 0°$ and increment it by 10° for a complete revolution. Find the angular position of link 4, the linear length of L3 (B_o to point C), and the vertical height of point D above the x-axis ($B_o - A_o$) assuming Figure 2.57 is drawn roughly to scale. Plot the vertical height of point D versus the input angle θ_2.

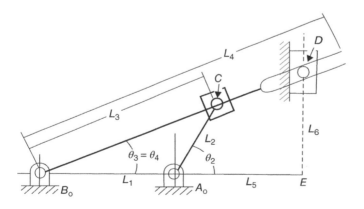

Figure 2.57   Programming Exercise P2.4

P2.5 Given: $L_1 = 3.00$ in., $L_2 = 1.38$ in., $L_3 = 2.50$ in., $L_4 = 2.75$ in., $L_5 = 4.75$ in., $L_6 = 4.88$ in., $L_7 = 3.25$ in., $\theta_1 = 180°$, $\theta_7 = -44°$, $L_{cp} = 2.88$ in. and $L_{pd} = 4.25$ in. Start with $\theta_2 = 0°$ and increment it by $5°$ for a complete revolution. Find the angular positions of links 3, 4, 5, and 6. Find the current location of point $P = (P_x, P_y)$ assuming Figure 2.58 is drawn roughly to scale. Plot P_y versus P_x to show the true shape of the coupler curve generated by point P. Origin at B_o.

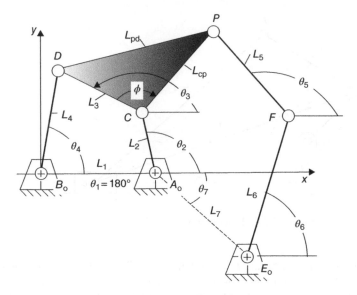

Figure 2.58 Programming Exercise P2.5

3

Graphical Design of Planar Linkages

3.1 Introduction

Design engineers are often faced with the task of creating a linkage that generates an irregular motion. Synthesis is the design of a linkage to produce a desired output motion for a given input motion. Chapter 2 acquainted the reader with the position analysis of various planar linkages. This chapter introduces the reader to the graphical design of planar linkages. It follows Chapter 2 because after you have designed a linkage, you need to be able to verify that it moves according to its motion design specifications.

History shows that graphical synthesis of linkages was predominately used before the age of the computer. It is included in this textbook because it helps the reader understand some of the basic concepts necessary for linkage synthesis, even if it is done using a computer. Graphical synthesis will typically get the design close to a possible solution or at least an insight into what might need to be changed to make a linkage design function properly.

Another reason for looking at graphical synthesis is that there are typically more unknown quantities in the design of a planar linkage than there are known physical conditions that must be met. This leads to an infinite number of solutions and an unsolvable solution for a computer program. A graphical synthesis can help the designer narrow down the number of unknowns graphically and through common sense and past experience. Design is typically an exercise in trade-offs between what you want and what you can obtain, while still designing a simple linkage.

A computer analysis results in a large set of numbers, which may be hard to interpolate. Converting some of these numbers to a graphical picture can provide the designer with valuable insight as to what might need to be modified to obtain the desired output motion.

Once a design concept has been determined, a graphical layout of the planar linkage can aid in the dimensional synthesis of the linkage. That is, the assigning of dimensions to the linkage.

Design and Analysis of Mechanisms: A Planar Approach, First Edition. Michael J. Rider.
© 2015 John Wiley & Sons, Ltd. Published 2015 by John Wiley & Sons, Ltd.

The desired output of a planar linkage might be for function generation, path generation, or motion generation. **Function generation** is defined as requiring a specific output for each specified input, sort of like a black box with an input and an output. **Path generation** is defined as the control of a specified point such that it follows a specified path in 2D space. The orientation of the point is not considered, just its location. **Motion generation** is defined as the control of a rigid body such that its location and orientation relative to a predefined coordinate system are defined. This concept shows up when it is desired to move a container from point A to B, then tilt it to empty its contents. Both its path and its orientation are important throughout the motion.

3.2   Two-Position Synthesis for a Four-Bar Linkage

First, we will discuss a 2-position synthesis of a 4-bar linkage. In this case, it is desired to have the output link oscillate between two extreme limits when being driven by an input link that rotates through 360° (driven by a motor). The output link will be at its extreme right limit when links 2 and 3 are inline as shown in Figure 3.1a. In this case, the distance from A_o to D is equal to the length of link 3 plus the length of link 2. Its extreme left position will occur when link 2 is folded over on link 3 as shown in Figure 3.1b. In this case, the distance from A_o to D is equal to the length of link 3 minus the length of link 2.

We can use this information to design a 4-bar linkage whose output link, link 4, oscillates between two specified angles while the input link, link 2, rotates through 360°.

Example 3.1   Design a 4-bar linkage with angular motion displacement of 55° for its output link

Problem: Add a dyad (two links) to the output link to create a 4-bar linkage whose input is driven by a motor and whose output oscillates between 53° and 108° with the clockwise motion time being equal to the counterclockwise motion time. The bearings A_o and B_o should not be further apart than 12 in.

Solution: Procedure follows.

1. Pick a location for bearing B_o, draw it, and then specify an x–y reference axis.
2. Draw the output link, link 4, in its two extreme positions, $\theta_{41} = 53°$ and $\theta_{42} = 108°$.
3. Draw a line segment between the ends of link 4 labeled D_1 and D_2, and then extend this line segment in either direction.
4. Bisect the line segment D_1D_2.

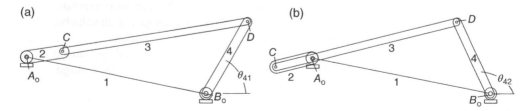

Figure 3.1   Extreme positions of 4-bar. (a) Links 2 and 3 extended and (b) links 2 and 3 overlap

5. Select a location on this line such that bearing A_o is less than the specified distance from bearing B_o as defined in the problem statement. Link 3 should be 3–4 times the length of link 2. If you make $(A_o$ to $D_2) = (D_2$ to $D_1)$, then L_3 is three times L_2.

6. Set your compass at a radius equal to half of line segment D_1D_2. Place your compass at bearing A_o and draw a circle. This will be the path of point C at the end of link 2. This distance from A_o to C is the length of link 2.

7. Label the right intersection point of the line and the circle as C_1. Label the left intersection point as C_2.

8. The distance between C_1 and D_1 is the same as the distance between C_2 and D_2. This distance is the length of link 3, the coupler link.

9. Measure the distances between A_o and C_1, C_1 and D_1, D_1 and B_o, and A_o and B_o. These four distances represent the lengths for link 2, link 3, link 4, and link 1 (see Figure 3.2).

10. Check the Grashof condition to be sure that the input link, link 2, can rotate through 360° without taking the linkage apart, then putting it back together again in its second configuration. Need: $L_{shortest} + L_{longest} < L_{other} + L_{last}$. If this is not true, select a new location for bearing A_o further from B_o and try again. If this is not possible, shorten link 4 and start again.

11. Use the information learned in Chapter 2 to analyze the linkage in its two extreme positions to verify your design.

Answers: $\boxed{L_1 = 10.0'' \text{ at } 138° \text{ from } B_o \text{ to } A_o}$, $\boxed{L_2 = 2.77''}$, $\boxed{L_3 = 8.47''}$, and $\boxed{L_4 = 6.0''}$.

Since C_2, A_o, and C_1 lie along the same line, the angle of rotation from C_1 to C_2 is 180° and from C_2 to C_1 is 180°. This guarantees that the clockwise motion time for link 4 will be equal to its counterclockwise motion time.

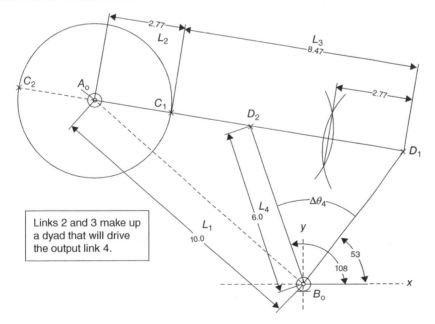

Figure 3.2   Graphical 2-angular positions

3.3 Two-Position Synthesis for a Quick Return 4-Bar Linkage

If you want to design a quick-return 4-bar linkage where the clockwise motion is longer than the counterclockwise motion, then the design procedure above needs to be modified slightly. Before we do this, let us look at a 4-bar linkage with an unequal timing ratio for its output oscillation (consider Figure 3.3). With the input link, link 2, rotating counterclockwise, the output link, link 4, will move from D_1 to D_2 after link 2 has rotated through 196.7°. Link 4 will move from D_2 back to D_1 after link 2 has rotated through 163.3°. If the input link is rotating at a constant speed, for example, driven by a motor, then link 4's timing clockwise will be different from its timing counterclockwise.

If the input is rotating at 1 revolution per second, then it will take $\dfrac{196.7}{360}(1 \text{ s}) = 0.546$ s to move from D_1 to D_2. It will take 0.454 s to return from D_2 to D_1, thus the quick return. Note that 196.7° minus 163.3° equals 33.4° or two times 16.7°. We can use this information when designing a quick-return 4-bar linkage. In addition, the timing ratio is $Q = \dfrac{196.7}{163.3} = \dfrac{0.5464}{0.4536} \approx 1.205$. The time over is approximately 20% higher than the return time. If you know the timing ratio, Q, then you can calculate the two input angles between the extreme positions and the angle, δ, between link 3 in its two extreme positions. For example, given a timing ratio of $Q = 1.2$, the construction angle delta can be calculated as follows.

$$\delta = 180 \cdot \left(\frac{Q-1}{Q+1}\right)$$

$$\delta = 180 \cdot \left(\frac{1.2-1}{1.2+1}\right) = 16.4°$$

$$\theta_{\text{long}} = 180° + \delta = 180° + 16.4° = 196.4°$$

$$\theta_{\text{short}} = 180° - \delta = 180° - 16.4° = 163.6°$$

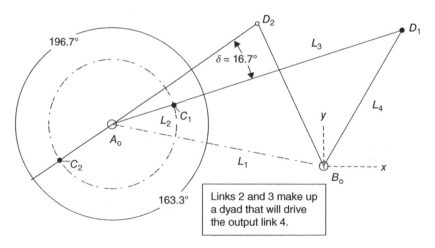

Figure 3.3 Quick-return 4-bar

These values come very close to the values shown in Figure 3.3. The difference lies in the fact that the timing ratio was slightly greater than 1.2 and we used 1.2 for these calculations.

Example 3.2   Design a quick-return 4-bar linkage with an angular motion displacement of 58° for its output link

Problem: Design a quick-return 4-bar linkage whose input is driven by a constant speed motor and whose output oscillates between 58° with the clockwise motion time being 1.25 times slower than the counterclockwise motion time. The bearings A_o and B_o should not be further apart than 12 in.

Solution: Procedure follows.

1. Pick a location for bearing B_o, draw it, and then specify an x–y reference axis.
2. Pick an initial length and angle for the output link, link 4. Draw link 4 in its two extreme positions, (I picked) $\theta_{41} = 52°$ and $\theta_{42} = 110°$. Label the two ends of link 4, D_1 and D_2.
3. Draw a line segment at a 45° angle from link 4, an unspecified length.
4. Using the timing ratio, Q, calculate the construction angle, δ.

$$\delta = 180° \cdot \left(\frac{Q-1}{Q+1}\right) = 180° \cdot \left(\frac{1.25-1}{1.25+1}\right) = 20°$$

5. Calculate the orientation of a second line so that it is oriented δ away from the first line and passes through D_2 (see Figure 3.4). $(\theta_{41} = 52°) - 45° + (\delta = 20°) = 27°$, so draw a line at 27° that passes through D_2.
6. The intersection of these two line segments is the location for bearing A_o. Be sure the location of A_o is less than the specified distance from bearing B_o as defined in the problem statement if appropriate.
7. The distance from A_o to D_1 is the sum of links 3 and 2. The distance from A_o to D_2 is the difference between links 3 and 2. Measure these two distances, and then calculate

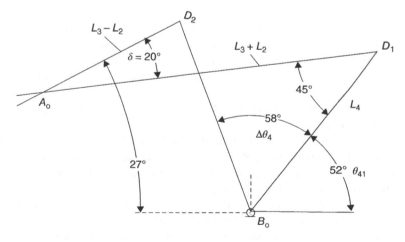

Figure 3.4   Construct second line

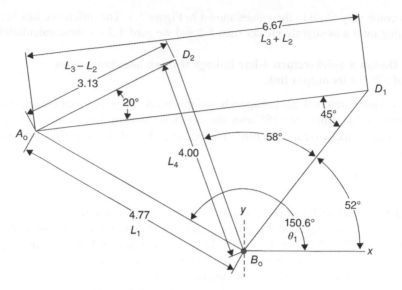

Figure 3.5 Quick-return 2-angular positions

the lengths of links 2 and 3 by solving two linear equations for two unknowns (see Figure 3.5).

$$L_3 + L_2 = 6.67$$

$$L_3 - L_2 = 3.13$$

$$2L_3 = 9.80$$

$$L_3 = 4.90 \text{ in.}$$

$$L_2 = 4.90 - 3.13 = 1.77 \text{ in.}$$

8. Set your compass at a radius equal to the length of link 2. Place your compass at bearing A_o and draw a circle. This will be the path of point C at the end of link 2.
9. Label the right intersection point of the line and the circle between A_o and D_1 as C_1.
10. Extend line D_2 to A_o so it crosses the left edge of the circle. Label this left intersection point as C_2.
11. Measure the distances between D_1 and B_o, and A_o and B_o. These two distances represent the lengths for link 4 and link 1.
12. Check the Grashof condition to be sure that the input link, link 2, can rotate through 360° without taking the linkage apart, then putting it back together again in its second configuration. Need: $L_{shortest} + L_{longest} < L_{other} + L_{last}$. If this is not true, select an angle different from 45° for the first line segment and try again.
13. Use the information learned in Chapter 2 to analyze the linkage in its two extreme positions to verify your design. Also, verify that the timing is correct per the design statement. Note that the input, link 2, must rotate clockwise for the clockwise motion of the output link to be the longer time.

Answers: $\boxed{L_1 = 4.77'' \text{ at } 150.6° \text{ from } B_o \text{ to } A_o}$, $\boxed{L_2 = 1.77''}$, $\boxed{L_3 = 4.90''}$, and $\boxed{L_4 = 4.00''}$.

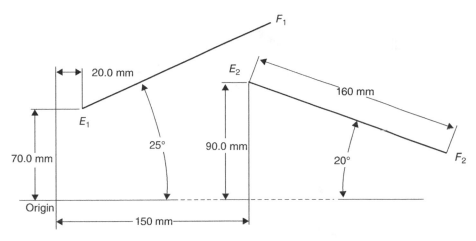

Figure 3.6   Output link in two positions

3.4   Two-Positions for Coupler Link

Next, we will look at the case where the output link contains a line segment that must move between two locations and orientations and be driven by a constant speed motor (see Figure 3.6).

Since this is part of link 4, the first step is to determine the pivot point for link 4 so that the defined segment will move and rotate as desired. This can be done by drawing a perpendicular bisector between E_1 and E_2 and a perpendicular bisector between F_1 and F_2. The intersection of these two bisectors is the proper location for the bearing, B_o, of link 4. Although the line B_oE_2 looks perpendicular to E_2F_2, it may not be. This angle on Figure 3.7 is 83.8°. Note that the angle $B_oE_1F_1$ is equal to the angle $B_oE_2F_2$.

Add a dyad to drive link 4 between these two extreme positions. The rest of the 4-bar linkage design is exactly like the previous two examples since the extreme limits of link 4 are now defined. Follow Example 3.1 if the timing ratio is one. Follow Example 3.2 if a quick-return 4-bar linkage is desired.

3.5   Three Positions of the Coupler Link

In the case where the coupler link (see Figure 3.8) must be in three different locations and at three different orientations, the following method can be used. Since C_1, C_2, and C_3 are equal distance from bearing A_o, drawing a circle through these three points will locate bearing A_o. In addition, D_1, D_2, and D_3 are equal distance from bearing B_o, and thus drawing a circle through these three points will locate bearing B_o.

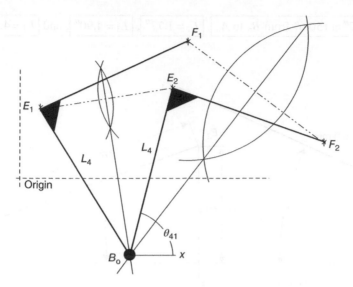

Figure 3.7   Locate bearing B_o

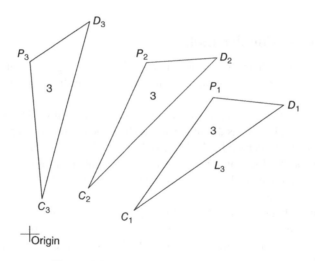

Figure 3.8   Coupler link in three positions

Example 3.3   Design a 4-bar linkage whose coupler link goes through three specific locations

Problem: Design a 4-bar linkage whose 5.00 in. long coupler link goes through the three positions shown in Figure 3.9, and then determine if the 4-bar's link 2 can rotate through 360°. Position 1: end point at (0.50, 3.38) inches and link 3 horizontal. Position 2: end point at (3.00, 3.84) inches and link 3 is 17° clockwise from horizontal. Position 3: end point at (3.75, 2.44) inches and link 3 is 26 clockwise from horizontal.

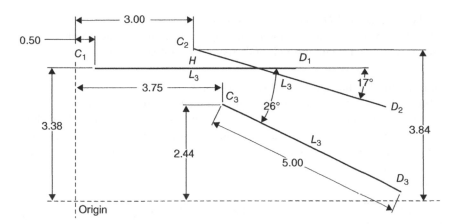

Figure 3.9 Three positions for coupler link

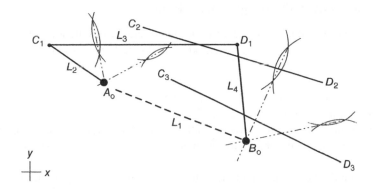

Figure 3.10 4-Bar designed

Solution: Procedure follows.

1. Draw the coupler link, link 3, in its three specified locations.
2. Mark one end of the coupler link as C_1, C_2, and C_3.
3. Mark the other end on link 3 as D_1, D_2, and D_3.
4. Draw a perpendicular bisector between points C_1 and C_2.
5. Draw a perpendicular bisector between points C_2 and C_3.
6. The intersection of these two perpendicular bisectors is the center for a circle that will go through points C_1, C_2, and C_3. This is the location for bearing A_o.
7. Draw a line between A_o and C_1. This will be link 2 (see Figure 3.10).
8. Draw a perpendicular bisector between points D_1 and D_2.
9. Draw a perpendicular bisector between points D_2 and D_3.
10. The intersection of these two perpendicular bisectors is the center for a circle that will go through points D_1, D_2, and D_3. This is the location for bearing B_o (see Figure 3.10).

11. Draw a line between B_o and D_1. This will be link 4.
12. Measure the length of Links 2 and 4. Locate the bearings relative to the origin.
13. Check to see if link 2 can rotate through 360° using Grashof condition.

Grashof Condition for this linkage: $(L_2 + L_3) < (L_1 + L_4)$?

Is $(1.78 + 5.00) < (4.08 + 2.57) \rightarrow$ NO, Link 2 will not rotate through 360°. This is a triple-rocker linkage. No link will rotate through 360°. If we want this linkage to move through the three positions defined, then we will have to add a dyad (two more links with one of these links rotating through 360°) to our design solution. We can select either link 2 or link 4, determine its range of motion, and then add links 5 and 6 with link 6 being driven by a motor (see Example 3.1).

Answer: $\boxed{L_2 = 1.78''}$, $\boxed{L_4 = 2.57''}$, $\boxed{L_1 = 4.08''}$, $\boxed{A_o = (1.97'', 2.39'')}$, $\boxed{B_o = (5.74'', 0.82'')}$, and $\boxed{L_3 = 5.00''}$.

3.6   Coupler Point Goes Through Three Points

If we need a coupler point to go through three different points and the orientation of the coupler link is not important, then a different design method is used (see Figure 3.11).

Procedure:

1. Locate the three prescribed positions (P_1, P_2, and P_3).
2. Select a location for the fixed pivot point, bearing A_o.
3. Choose a length for link 2 (A_oC) and draw the path of point C (circle) centered at A_o. Make sure the length of link 2 is greater than half the distance between the any two specified "P" points.
4. Pick a point for C_3 on the newly drawn circle, relative to P_3. Picking the point farthest from bearing A_o guarantees that link 3 should reach a possible location of link 2.
5. With length C_3P_3 determined and fixed, find C_2 from P_2 and C_1 from P_1. If the length C_2P_2 or C_1P_1 does not intersect the "C" circle drawn, then begin again, choosing a different location for C_3, thus a different length.

Figure 3.11   Three coupler points

6. Draw line C_1D_1 at some length and some angle away from C_1P_1. Draw line P_1D_1 to complete link 3.
7. Locate D_2 using the sizes from step 6 above. Link 3 is the same size in each position; that is, $C_1D_1 = C_2D_2 = C_3D_3$ and $P_1D_1 = P_2D_2 = P_3D_3$.
8. Locate D_3 using the sizes from step 6 above.
9. Construct perpendicular bisectors for D_1D_2, and D_2D_3 to locate the second bearing B_o as before. If you do not like the location of bearing Bo or the size of link 4, go back to step 2, 3, or 4 and begin again.
10. Measure the length of link 2, link 3, and link 4. Locate the position of A_o and B_o. The 4-bar linkage consists of:

$$L_1 = A_oB_o$$
$$L_2 = A_oC_1 = A_oC_2 = A_oC_3$$
$$L_3 = C_1D_1 = C_2D_2 = C_3D_3$$
$$L_4 = B_oD_1 = B_oD_2 = B_oD_3$$
$$CP = C_1P_1 = C_2P_2 = C_3P_3$$
$$DP = D_1P_1 = D_2P_2 = D_3P_3$$

11. Measure the three orientation angles for link 2 (θ_{21}, θ_{22}, and θ_{23}) so that you can verify that point P goes through the three prescribed points (P_1, P_2, and P_3).
12. Draw the mechanism in all three positions to see if it goes through the prescribed points P_1, P_2, and P_3. (Always check your work!)

Example 3.4   Design a 4-bar linkage whose coupler point goes through three specified locations

Problem: Design a 4-bar linkage whose input is driven by a motor and whose coupler point goes through three points located at $P_1 = (5.00'', 3.38'')$, $P_2 = (6.50'', 2.38'')$, and $P_3 = (6.68'', 1.25'')$.

Solution: The procedure is listed above. After locating the three points labeled P_1, P_2, and P_3 in your design area, select a location for bearing A_o. Try to pick a location that makes sense relative to the specified points. For this example, bearing A_o is located at $(1.00'', 0.75'')$ (see Figure 3.12).

Figure 3.13 shows the design after steps 3, 4, and 5 are completed. In the design 1, the change in link 2's angular position is very small when the coupler point moves from P_2 to P_3. This will cause a very quick change in location of the coupler and generate high inertia forces in the coupler link. Design 2 has the coupler link moving from P_1 to P_2 in about the same amount of time that it takes to move it from P_2 to P_3. Remember that link 2 is rotating at a constant rate. Design 2 is a better design since a quick motion between points P_2 and P_3 is not specified.

In step 6, we randomly create the size of the coupler link, link 3. There are an infinite number of solutions here so if the one you pick does not work well, come back to this step and try a different shape.

Next, we randomly select the length of link 3 from C_1 to D_1 and its orientation from the C_1P_1 line. Then duplicate this information to locate D_2 and D_3 (see Figure 3.14).

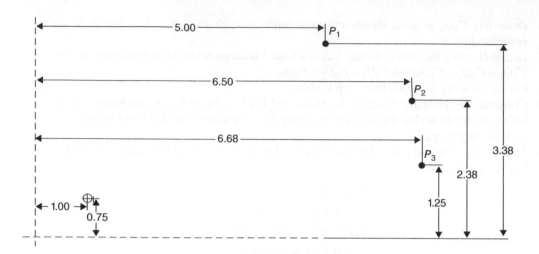

Figure 3.12 Three points and bearing A_o

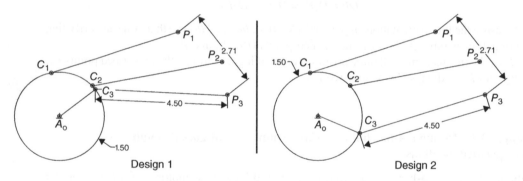

Design 1   Design 2

Figure 3.13 Two possible initial designs

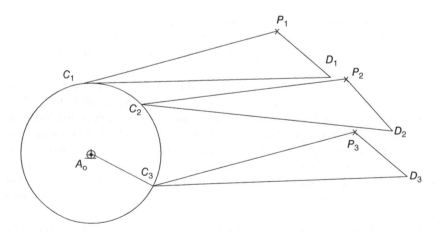

Figure 3.14 Adding the coupler link

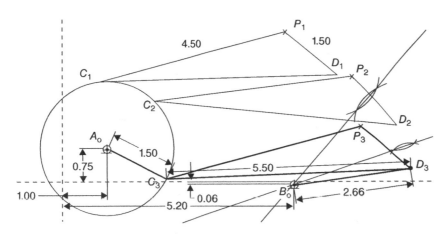

Figure 3.15   Final layout of linkage

By bisecting the imaginary lines between D_1 and D_2 and between D_2 and D_3, we can locate a point that is equal distance from all three "D" points. This is the location for bearing B_o. The final step in the layout is to determine the lengths of the links and the locations of the bearings (see Figure 3.15).

Answers: $A_o = (1.00, 0.75)$, $B_o = (5.20, -0.06)$, $L_2 = 1.50$, $L_3 = 5.50$, $L_4 = 2.66$, $CP = 4.5$, and $DP = 1.5$ in.

Although the design layout is complete, there are still some issues to address. For example, the transmission angle between the coupler link, link 3, and the output link, link 4, is very small when the linkage is in the third position. This will cause excessive forces, thus this is a bad design. Checking Grashof's condition shows that the input link, link 2, cannot rotate through a complete 360° cycle, and thus a dyad needs to be added to force link 2 to oscillate between its two extreme positions C_1 and C_3.

$$L_1 = \sqrt{(B_{ox} - A_{ox})^2 + (B_{oy} - A_{oy})^2} = 4.25 \text{ in.}$$

shortest + longest < other_two_links

$1.50 + 5.50 < 2.66 + 4.25$ (NOT True!!)

A second attempt at a linkage design leads to the following, Figure 3.16. Bearing $A_o = (1.80, 5.00)$, $B_o = (4.86, 1.54)$, $L_1 = 4.62$, $L_2 = 1.50$, $L_3 = 4.80$, $L_4 = 1.88$, $CP = 4.80$, and $DP = 0.00$. The minimum transmission angle is slightly less than 40°. Grashof's condition states that link 2 will rotate through 360°. This is an acceptable design. Note that point P was selected as one end of the coupler link, link 3. There are many other acceptable designs for this problem.

3.7   Coupler Point Goes Through Three Points with Fixed Pivots and Timing

There are cases where the coupler point must go through three specified points along with some prescribed timing. That is, the input link must move through 40° as the coupler point moves

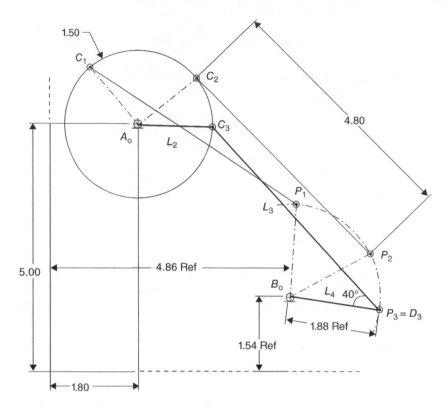

Figure 3.16 Acceptable design through three points

from point 1 to point 2 and through 90° as the coupler point moves from point 1 to point 3. Because of certain restrictions, the fixed bearings must be located in a specified location as well. If these conditions are present, then the following procedure can be used to design a 4-bar linkage that meets all the requirements. The procedure is outlined in the following example (see Figure 3.17).

Example 3.5 Design a 4-bar linkage whose coupler link goes through three specific points with prescribed timing and fixed pivot points

Problem: Design a 4-bar linkage whose coupler point goes through the following three points: $P_1 = (72, 147)$ mm, $P_2 = (128, 152)$ mm, and $P_3 = (163, 132)$ mm. In addition, the input crank must rotate through 40° clockwise as the coupler point moves from point 1 to point 2 and through 90° clockwise as the coupler point goes from point 1 to point 3. The bearing for the input crank must be located at the origin at $(0, 0)$ mm. The output link's bearing, B_o, must be located approximately 172 mm and 10° above the horizontal from bearing A_o.

Solution: Procedure follows.

1. Locate the three prescribed positions (P_1, P_2, and P_3).
2. Locate the fixed pivot points, bearing A_o and B_o.
3. Draw a construction line from P_2 to A_o.

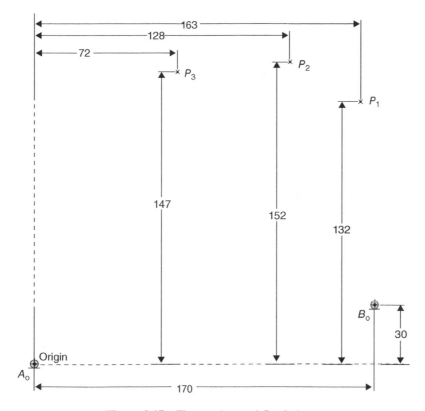

Figure 3.17   Three points and fixed pivots

4. We will be inverting the motion to fix the length of the unknown input link. Draw a second same length construction line rotated 40° counterclockwise (opposite direction from specification) about bearing A_o. Label its endpoint as P_2'.
5. Draw a construction line from P_3 to A_o.
6. Draw a second same length construction line rotated 90° counterclockwise (opposite direction from specification) about bearing A_o. Label its endpoint as P_3'.
7. Construct a perpendicular bisector between P_1 and P_2'.
8. Construct a perpendicular bisector between P_1 and P_3'.
9. The intersection of these two bisectors is the end of the crank, link 2. Label this point as C_1. The distance between C_1 and A_o is link 2's length. Draw a circle around bearing A_o to represent the path of point C (see Figure 3.18).
10. With the distance between C_1 and P_1 fixed, locate C_2 the same distance from P_2, and C_3 the same distance from P_3. Note that C_3 can be in two different locations on the circle and be the correct length. However, only one of those locations is 90° from link 2 in its original position at C_1 (see Figure 3.19).
11. Now we are ready to deal with bearing B_o.
12. The position of D is found by means of kinematic inversion. This is done by fixing the coupler link in position 1. The rest of the mechanism including the frame must move so that the same relative motion exists between all links in this inversion as well as the

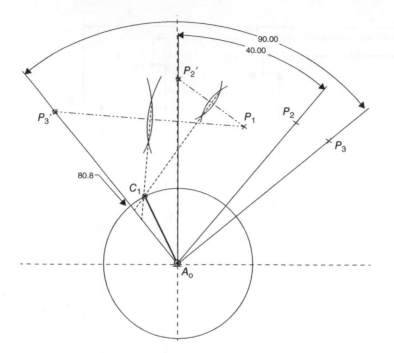

Figure 3.18   Locate end of crank link, C_1

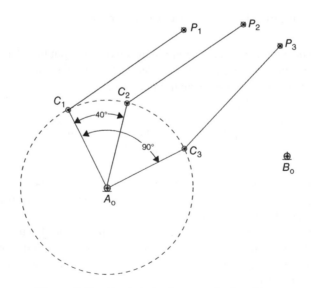

Figure 3.19   Crank in its three required positions

original configuration. The relative positions of B_o with respect to position 1 of the coupler are obtained by construction.

13. Rotate A_o about C_1 by $(\beta_2 - \beta_1)$ where $\beta_1 = \angle A_o C_1 P_1$ and $\beta_2 = \angle A_o C_2 P_2$. Label the endpoint A_o'.
14. Draw an arc around A_o' with a radius equal to the length of link 1, or $\overline{A_o B_o}$.
15. Draw an arc around P_1 with a radius equal to $\overline{P_2 B_o}$. The intersection of these two arcs is the location of B_o'.
16. Rotate A_o about C_1 by $(\beta_3 - \beta_1)$ where $\beta_1 = \angle A_o C_1 P_1$ and $\beta_3 = \angle A_o C_3 P_3$. Label the endpoint A_o''.
17. Draw an arc around A_o'' with a radius equal to the length of link 1, or $\overline{A_o B_o}$.
18. Draw an arc around P_1 with a radius equal to $\overline{P_3 B_o}$. The intersection of these two arcs is the location of B_o''.
19. Construct perpendicular bisector to the lines $B_o B_o'$ and $B_o' B_o''$. The intersection locates D_1, which is the other end of the coupler link.
20. Draw the linkage in all three positions to check your design. If the design is not acceptable, these steps can be repeated with a different location for bearings A_o and B_o.

Answers: | All graphical designs will be different |.

3.8   Two-Position Synthesis of Slider-Crank Mechanism

The inline slider-crank mechanism shown in Figure 3.20 has a stroke equal to twice the crank length. It has a forward stroke time equal to its reverse stroke time for a constant rotational speed of the crank. The extreme positions D_1 and D_2 are also known as the limiting positions. In general, the connecting link, link 3, must be larger than the crank, link 2. Good design practices assume the connecting link to three to four times the crank length. Shorter lengths of the connecting link gives rise to higher velocity and acceleration vectors for link 3.

In Figure 3.20, the desired stroke is 5.00 in. and therefore the crank is 2.50 in. and the connecting link is 7.50 in.

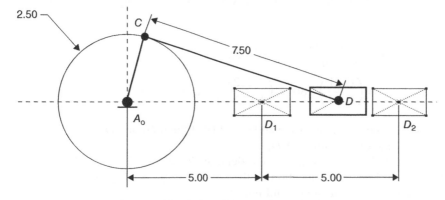

Figure 3.20   Inline slider-crank mechanism

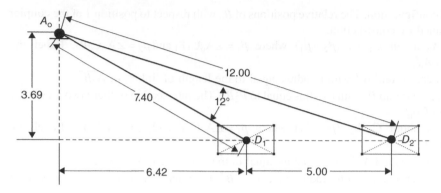

Figure 3.21 Offset slider-crank mechanism

The offset slider-crank mechanism shown in Figure 3.21 has a different time for its forward and reverse strokes if the crank is rotating at a constant speed. This feature can be used to synthesis a quick return mechanism where a slower working stroke is desired. In addition, the stroke D_1D_2 is always greater than twice the crank length. Once again, the connecting link should be greater than three times the crank length.

The offset slider-crank in Figure 3.21 will create a forward motion of the slider for 0.08 s and a reverse motion for 0.07 s if the crank makes a complete rotation in 0.15 s. The slider's stroke was specified as 5.00 in. The crank will be $(12.00 - 7.40)/2 = 2.30$ in. The connecting link will be $(12.00 - 2.30) = 9.70$ in. The crank's bearing will be 3.69 in. above the line of action of the slider and 6.42 in. to the left of the extreme left position of the slider.

Example 3.6 Design a slider-crank mechanism that will move the slider to the right slower than it moves it to the left

Problem: Design a slider-crank mechanism that will move the slider to the right in 1.1 s and return in approximately 0.90 s with a stroke of 6.00 in.

Solution: The procedure follows.

1. Determine the timing ratio and then the connector link's delta angle between the two extreme limits of the slider.

$$Q = \frac{\text{slow\_time}}{\text{fast\_time}} = \frac{1.1}{0.9} = 1.222$$

$$\beta = \left(\frac{Q-1}{Q+1}\right) \cdot 180° = \left(\frac{1.222-1}{1.222+1}\right) \cdot 180° = 18.0°$$

2. Locate the two extreme positions of the slider and label them D_1 and D_2.
3. Draw a random slanted line starting at D_1.
4. Draw a line through D_2 that has the correct delta angle calculated in step 1. The intersection of these two lines is the location of the fixed bearing, A_o. If you do not like the location of A_o, then go back to step 3 and repeat.
5. Measure the distance between A_o and D_1.

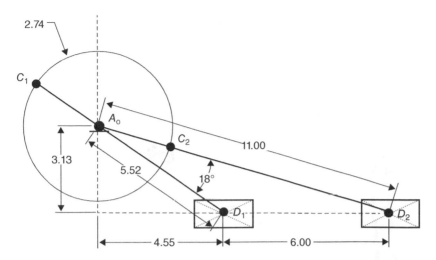

Figure 3.22 Offset slider-crank design

6. Measure the distance between A_o and D_2.
7. Calculate the crank length and the connecting link's length. If you do not like the values generated, then go back to step 3 and repeat.
8. Locate the A_o relative to one of the extreme positions of the slider (see Figure 3.22).
9. Construct the slider-crank mechanism and verify the design.

$$L_2 = \frac{11.00 - 5.52}{2} = 2.74''$$
$$L_3 = 11.00 - 2.74 = 8.26'' \, (> 3L_2)$$

Answers: $\boxed{L_1 = 3.13''}$, $\boxed{L_2 = \text{crank} = 2.74''}$, $\boxed{L_3 = \text{connecting link} = 8.26''}$, and $\boxed{X_{A_oD1} = 4.55 \text{ in.}}$

3.9   Designing a Crank-shaper Mechanism

The crank-shaping mechanism is used to machine flat metal surfaces especially where a large amount of metal has to be removed. The reciprocating motion of the mechanism inside the crank-shaping mechanism can be seen in Figure 3.23. As the input link (disc) rotates, the top of the machine moves forward and backward, pushing a cutting tool. The cutting tool removes the material from the work piece, which is held firmly.

The crank-shaping machine is a simple and yet extremely effective mechanism. It is used to remove material, usually steel or aluminum, to produce a flat surface. However, it can also be used to manufacture gear racks and other complex shapes.

Figure 3.24 shows the kinematic diagram for a crank-shaper. Link 2 is the input crank and link 6 is the tool cutter.

Procedure assuming the slider, link 6, moves horizontally:

1. Draw the slider in its two extreme positions and label them F_1 and F_2.

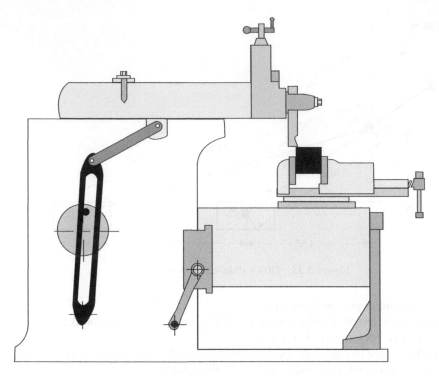

Figure 3.23 Crank-shaper machine

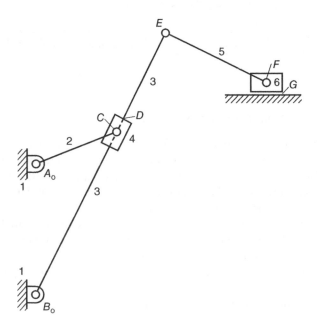

Figure 3.24 Kinematic diagram for crank-shaper

2. Draw link 5 at some angle and pick its length. Label its endpoints E_1 and E_2. Maximum force can be transmitted to the slider, link 6, if link 5 is nearly horizontal in this step.
3. Calculate the timing ratio, Q, as before.

$$Q = \frac{\text{Slower stroke time}}{\text{Faster stroke time}} \geq 1.00$$

4. Calculate the sweep angle for link 3.

$$\beta = \left(\frac{Q-1}{Q+1}\right)180°$$

5. Draw a line going through E_1 at an angle of $(-90° - \frac{1}{2}\beta)$.
6. Draw a line going through E_2 at an angle of $(-90° + \frac{1}{2}\beta)$.
7. To check your work, draw a perpendicular bisector through the imaginary line between E_1 and E_2. It should intersect at the same location where steps 5 and 6 above meet. This intersection point is fixed bearing B_0.
8. Locate on this vertical perpendicular bisector a point where you want to place fixed bearing A_0. The distance from B_0 to A_0 must be greater than the length of the crank, link 2. If it is not, move farther up the perpendicular bisector when locating A_0.
9. Calculate the crank-shaper mechanism link sizes (see Figure 3.25).

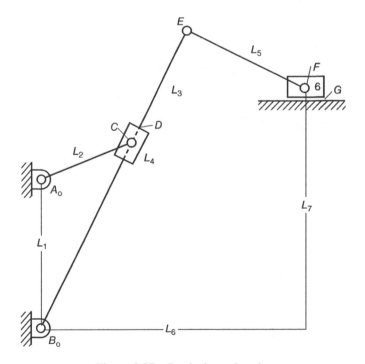

Figure 3.25   Crank-shaper lengths

$$L_1 = \overline{A_oB_o} \quad L_2 = L_1 \sin\left(\frac{1}{2}\beta\right)$$

$$L_3 = B_oE_1 = B_oE_2$$

$$L_4 = B_oC$$

$$L_5 = E_1F_1 = E_2F_2$$

$$L_6 = \text{horizontal distance from } B_o \text{ to } F$$

$$L_7 = \text{vertical distance from } B_o \text{ to } F$$

Example 3.7   Design a crank-shaper with a specified timing ratio and stroke length

Problem: Design a crank-shaper with a timing ratio of 1.2 and stroke length of 150 mm. The slower stroke takes 0.140 s as the slider moves from left to right. What is the input crank speed? What are the link lengths as defined in the procedure above? Assume the slider, link 6, moves horizontal.

Solution:

1. Draw the slider in its two extreme positions and label them F_1 and F_2.
2. Draw link 5, 5° from horizontal, and pick its length to be 200 mm. Label the endpoints E_1 and E_2.
3. Timing ratio was given as 1.20.

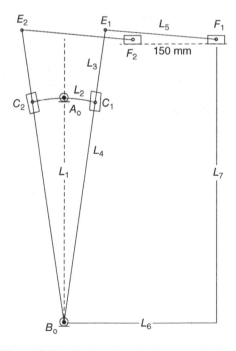

Figure 3.26   Designed crank-shaper mechanism

4. Calculate the sweep angle for link 3. $B = 16.4°$.
5. Draw a line going through E_1 at an angle of $-98.2°$.
6. Draw a line going through E_2 at an angle of $-81.8°$.
7. Construct perpendicular bisector to check your accuracy, thus locating fixed bearing B_o.
8. Locate A_o on the perpendicular bisector.
9. Determine all of the vector lengths for the crank-shaper (see Figure 3.26).

Slow stroke goes through $196.4°$ and fast stroke goes through $163.6°$, thus $Q = 1.2$.

Complete cycle happens in 0.2566 s, thus the $\boxed{\text{crank rotates at } 234 \text{ rpm.c.w.}}$

Answers: $\boxed{L_7 = 503 \text{ mm}}$, $\boxed{L_5 = 200 \text{ mm}}$, $\boxed{L_3 = 526 \text{ mm}}$, $\boxed{L_1 = 400 \text{ mm}}$, and $\boxed{L_2 = 57 \text{ mm}}$.

Problems

3.1   Design a 4-bar linkage that will cause the 3.00-in. output link to oscillate through $72°$ with equal forward and backward motion.
3.2   Design a 4-bar linkage that will cause the 80-mm output link to oscillate through $72°$ with equal forward and backward motion.
3.3   Design a 4-bar linkage that will cause the 2.50-in. output link to oscillate through $60°$. The clockwise motion should take 0.5 s while the counterclockwise motion should take 0.6 s. What is the required angular velocity vector of the input crank?
3.4   Design a 4-bar linkage that will cause the 50-mm output link to oscillate through $60°$. The clockwise motion should take 1.2 s while the counterclockwise motion should take 1.0 s. What is the required angular velocity vector of the input crank?
3.5   Design a 4-bar linkage that will cause the 3.50-in. output link to oscillate through $48°$. The clockwise motion should take 0.4 s while the counterclockwise motion should take 0.5 s. What is the required angular velocity vector of the input crank?
3.6   Design a 4-bar linkage that will cause the 90-mm output link to oscillate through $48°$. The clockwise motion should take 0.5 s while the counterclockwise motion should take 0.4 s. What is the required angular velocity vector of the input crank?
3.7   The output link must oscillate between the two positions shown in Figure 3.27. Determine the required pivot point, and then add a dyad so a motor with constant rotation can move this link between its two extreme positions with a cycle time of 0.75 s.
3.8   The output link must oscillate between the two positions shown in Figure 3.28. Determine the required pivot point and then add a dyad so a motor with constant rotation can move this link between its two extreme positions with a cycle time of 0.80 s.
3.9   Design a 4-bar linkage that will move the coupler link through the three positions shown in Figure 3.29. Can an electric continuously rotating motor drive this linkage or must a dyad be added to this linkage? The coupler link must travel through the three positions shown, but it does not have to be limited to this range.
3.10   Design a 4-bar linkage that will move the coupler link through the three positions shown in Figure 3.30. Can this linkage be driven by an electric continuously-rotating motor or must a dyad be added to this linkage? The coupler link must travel through the three positions shown, but it does not have to be limited to this range.

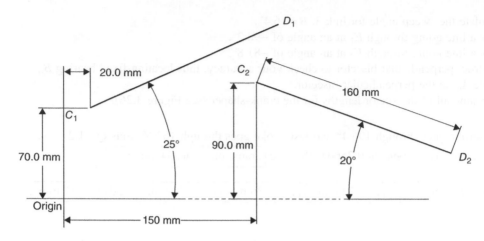

Figure 3.27   Problem 3.7

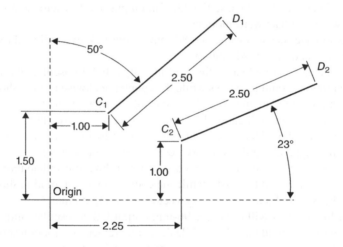

Figure 3.28   Problem 3.8

3.11   Design a 4-bar linkage that will move the coupler link through the three positions shown in Figure 3.31. Can this linkage be driven by an electric continuously-rotating motor or must a dyad be added to this linkage? The coupler link must travel through the three locations shown, but it does not have to be limited to this range.

3.12   Design a 4-bar linkage that will move the coupler link through the three positions shown in Figure 3.32. Can this linkage be driven by an electric continuously-rotating motor or must a dyad be added to this linkage? The coupler link must travel through the three locations, but it does not have to be limited to this range.

3.13   Design a 4-bar linkage that will move the coupler link through the three positions shown in Figure 3.33. Can this linkage be driven by an electric continuously-rotating motor or must a dyad be added to this linkage? The coupler link must travel through the three locations shown, but it does not have to be limited to this range.

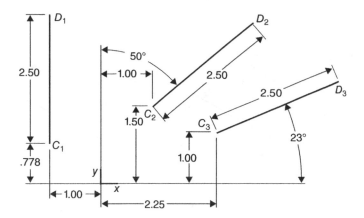

Figure 3.29   Problem 3.9

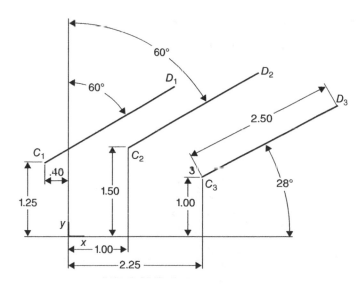

Figure 3.30   Problem 3.10

3.14   Design a 4-bar linkage that will move the coupler link through the three positions shown in Figure 3.34. Can this linkage be driven by an electric continuously-rotating motor or must a dyad be added to this linkage? The coupler link must travel through the three locations shown, but it does not have to be limited to this range.

3.15   Design a slider-crank mechanism with equal forward and reverse strokes and having a stroke length of 4.00 in.

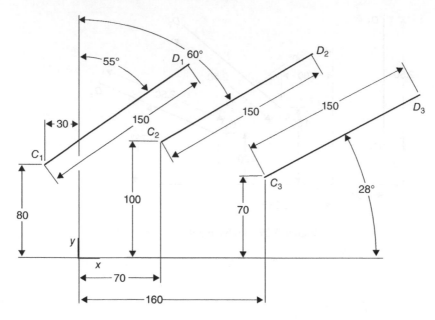

Figure 3.31   Problem 3.11

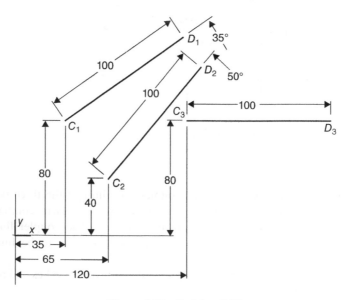

Figure 3.32   Problem 3.12

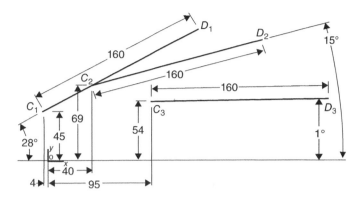

Figure 3.33   Problem 3.13

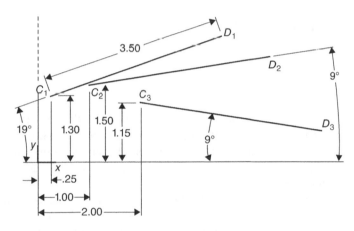

Figure 3.34   Problem 3.14

3.16   Design a slider-crank mechanism with equal forward and reverse strokes and having a
       stroke length of 150 mm.

3.17   Design an offset slider-crank mechanism with a stroke length of 4.00 in. and a timing
       ratio of 1.15.

3.18   Design an offset slider-crank mechanism with a stroke length of 150 mm and a timing
       ratio of 1.20.

3.19   Design a 4-bar linkage that goes through three precision points with a fixed pivot for link
       2 at the origin. $P_1 = (5.5, 85)$ mm, $P_2 = (24, 91)$ mm, and $P_3 = (68, 85)$ mm. Let link 2 be
       equal to 50 mm. After designing the 4-bar, be sure to define all variables shown in
       Figure 3.35. What are the three values for θ_2 so that the 4-bar's point P is located at
       P_1, P_2, and P_3?

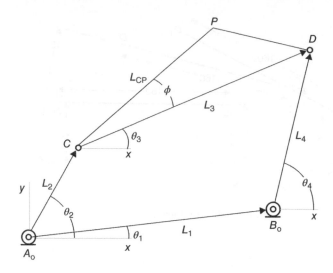

Figure 3.35   Problems 3.19 and 3.23 concept

3.20   Design a 4-bar linkage that goes through three precision points with a fixed pivot for link 2 at the origin. $P_1 = (0.0, 3^1/_8)$ inches, $P_2 = (2.0, 3\frac{3}{4})$ inches, and $P_3 = (3.0, 3^3/_8)$ inches. Let link 2 be equal to 2.5 in. After designing the 4-bar, be sure to define all variables shown in Figure 3.35. What are the three values for θ_2 so that the 4-bar's point P is located at P_1, P_2, and P_3?

3.21   Design a 4-bar linkage that goes through the three precision points. The locations for the fixed bearing A_o and B_o can be anywhere. $P_1 = (0, 55)$ mm, $P_2 = (20, 65)$ mm, and $P_3 = (40, 75)$ mm. After designing the 4-bar, be sure to define all variables shown in Figure 3.35. What are the three values for θ_2 so that the 4-bar's point P is located at P_1, P_2, and P_3?

3.22   Design a 4-bar linkage that goes through the three precision points. The locations for the fixed bearing A_o and B_o can be anywhere. $P_1 = (1, 3\frac{1}{2})$ inches, $P_2 = (2\frac{1}{2}, 4)$ inches, and $P_3 = (4, 3\frac{3}{4})$ inches. After designing the 4-bar, be sure to define all variables shown in Figure 3.35. What are the three values for θ_2 so that the 4-bar's point P is located at P_1, P_2, and P_3?

3.23   Design a crank-shaper mechanism with a timing ratio of 1.5 and a stroke length of 4.00 in. (see Figure 3.36).

3.24   Design a crank-shaper mechanism with a timing ratio of 1.4 and a stroke length of 100 mm (see Figure 3.36).

3.25   Design a crank-shaper mechanism with the slower stroke going left to right and taking 0.25 s. The return stroke takes 0.15 s. Its stroke length must be 6.00 in. What is the speed and direction of the input crank? (see Figure 3.36).

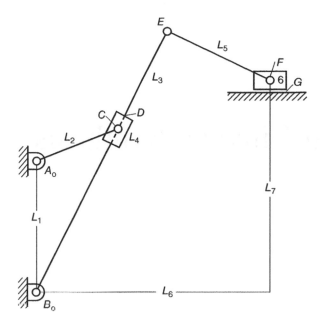

Figure 3.36   Crank-shaper lengths

4

Analytical Linkage Synthesis

4.1 Introduction

In this chapter, we will describe several methods for designing a planar 4-bar linkage and crank-slider mechanism based on the mechanism having certain requirements, such as function generation. This will be done using the Freudenstein's equation and Chebyshev spacing where appropriate.

4.2 Chebyshev Spacing

Before describing function generation synthesis, the optimal spacing for precision points will be presented. Chebyshev determined the best precision points to use when trying to curve fit a function using only a selected number of points. Chebyshev's polynomial method minimizes the error at the precision points and at both ends of the range and it is often used as a "best" first guess in the design of linkages. After the design is determined, slight modifications to these precision points can be done to optimize the design thus improving the mechanism's accuracy.

A simple construction is shown in Figure 4.1 for determining Chebyshev spacing as your initial guess. The precision points may be determined graphically on a circle whose diameter is proportional to the range of the independent parameter, Δx, when trying to curve fit the function $y = f(x)$. Assume you want to create "n" precision points. Draw a regular "$2n$-sided" equilateral polygon inscribed in a circle whose left and right edges are the limits of the independent variable "x." The left and right sides of the polygon must be drawn as vertical lines. Vertical lines drawn through the corners of the polygon intersect the horizontal diameter at the precision points.

Design and Analysis of Mechanisms: A Planar Approach, First Edition. Michael J. Rider.
© 2015 John Wiley & Sons, Ltd. Published 2015 by John Wiley & Sons, Ltd.

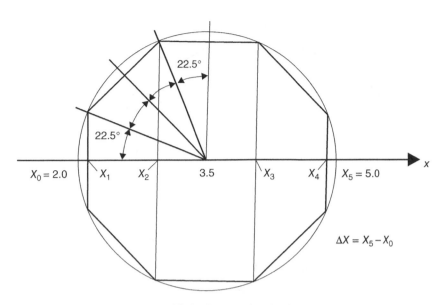

Figure 4.1   Chebyshev spacing for four points

Example 4.1   Determine the necessary precision points that should be used to design a function generating linkage given a specific function and the variable range

Problem: Determine three precision points for $(2 \leq x \leq 5)$ and the function $y = f(x) = x^2 - 2$.

Solution: Draw a circle with a diameter of 3.00 in., centered at 3.50 in. The left edge of this circle will be at 2.00 and the right edge will be at 5.00 in. Since three precision points are desired, a six-sided hexagon will be inscribed in this circle. Vertical lines through the corners of this polygon will intersect the horizontal diameter line at the precision points (see Figure 4.2).

Answers: $\boxed{x_1 = 2.20}$, $\boxed{x_2 = 3.50}$, and $\boxed{x_3 = 4.80}$.

Now using these x-values, determine the appropriate y-values based on the function given.

Answers: $\boxed{y_1 = f(x_1) = 2.84}$, $\boxed{y_2 = f(x_2) = 10.25}$, and $\boxed{y_3 = f(x_3) = 21.04}$.

The Chebyshev spacing can also be determined analytically as follows assuming "n" is the number of precision points desired, x_{min} is the lower limit of the independent variable, and x_{max} is the upper limit of the independent variable.

$$x_j = x_{min} + \frac{x_{max} - x_{min}}{2}\left[1 - \cos\left(\frac{(2j-1)\cdot(90°)}{n}\right)\right], \quad j = 1, 2, \ldots, n$$

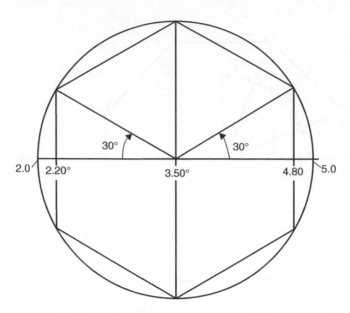

Figure 4.2   Chebyshev spacing for three points

Thus

$$x_1 = 2.00 + \frac{5.00-2.00}{2}\left[1-\cos\left(\frac{(2-1)\cdot 90°}{3}\right)\right] = 2.201$$

$$x_2 = 2.00 + \frac{5.00-2.00}{2}\left[1-\cos\left(\frac{(4-1)\cdot 90°}{3}\right)\right] = 3.500$$

$$x_3 = 2.00 + \frac{5.00-2.00}{2}\left[1-\cos\left(\frac{(6-1)\cdot 90°}{3}\right)\right] = 4.799$$

Example 4.2

Problem: Determine the Chebyshev spacing for four precision points on an input link with a starting angle is 40°, a range of 80°.

Solution: $\theta_{2min} = 40°$, $\theta_{2max} = 120°$, $n = 4$

$$\theta_{2j} = \theta_{2min} + \frac{\theta_{2max}-\theta_{2min}}{2}\left[1-\cos\left(\frac{(2j-1)\cdot(90°)}{4}\right)\right], \quad j = 1,2,3,4$$

Thus

$$\theta_{21} = 40° + \frac{120°-40°}{2}\left[1-\cos\left(\frac{(2-1)\cdot(90°)}{4}\right)\right] = 43.04°$$

$$\theta_{22} = 40° + \frac{120°-40°}{2}\left[1-\cos\left(\frac{(4-1)\cdot(90°)}{4}\right)\right] = 64.69°$$

$$\theta_{23} = 40° + \frac{120° - 40°}{2}\left[1 - \cos\left(\frac{(6-1)\cdot(90°)}{4}\right)\right] = 95.31°$$

$$\theta_{24} = 40° + \frac{120° - 40°}{2}\left[1 - \cos\left(\frac{(8-1)\cdot(90°)}{4}\right)\right] = 116.96°$$

Answers: $\boxed{\theta_{21} = 43.0°}$, $\boxed{\theta_{22} = 64.7°}$, $\boxed{\theta_{23} = 95.3°}$, and $\boxed{\theta_{24} = 117.0°}$.

4.3   Function Generation Using a 4-Bar Linkage

In order to design the 4-bar linkage we need a relationship between θ_2 and θ_4, thus eliminating θ_3 from the vector equations.

Figure 4.3 shows a 4-bar linkage and the vector loop needed for a position analysis. The vector loop equation is:

$$\overline{L_2} + \overline{L_3} - \overline{L_4} - \overline{L_1} = 0$$

This can be broken up into its y-components and its x-components. Note that $\theta_1 = 0$ so that L_1 only appears in the x-component equation.

$$y: \ L_2 \sin\theta_2 + L_3 \sin\theta_3 - L_4 \sin\theta_4 = 0$$

$$x: \ L_2 \cos\theta_2 + L_3 \cos\theta_3 - L_4 \cos\theta_4 - L_1 = 0$$

The next step is to eliminate θ_3 from the two equations by placing θ_3 on one side of the equation and everything else on the right side, then squaring each equation and adding the two equations together to form one equation without θ_3. Note that $\sin^2\theta + \cos^2\theta = 1$.

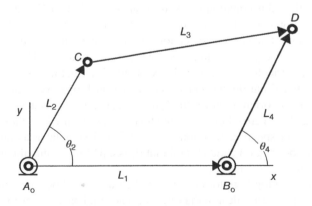

Figure 4.3   Four-bar linkage and vector loop

$$L_3 \sin\theta_3 = L_4 \sin\theta_4 - L_2 \sin\theta_2$$

$$L_3 \cos\theta_3 = L_4 \cos\theta_4 - L_2 \cos\theta_2 + L_1$$

$$L_3{}^2 \sin^2\theta_3 = L_4{}^2 \sin^2\theta_4 + L_2{}^2 \sin^2\theta_2 - 2L_4 L_2 \sin\theta_4 \sin\theta_2$$

$$L_3{}^2 \cos^2\theta_3 = L_4{}^2 \cos^2\theta_4 + L_2{}^2 \cos^2\theta_2 + L_1{}^2 - 2L_4 L_2 \cos\theta_4 \cos\theta_2 + 2L_4 L_1 \cos\theta_4 - 2L_1 L_2 \cos\theta_2$$

$$L_3{}^2 = L_4{}^2 + L_2{}^2 + L_1{}^2 - 2L_4 L_2 \sin\theta_4 \sin\theta_2 - 2L_4 L_2 \cos\theta_4 \cos\theta_2 + 2L_4 L_1 \cos\theta_4 - 2L_1 L_2 \cos\theta_2$$

Now divide both sides by $2L_2 L_4$ and rearrange the terms that leads to the following equation that only contains θ_4 and θ_2. This is Freudenstein's equation. It contains the relationship between the input rotation, θ_2, and the output rotation, θ_4.

$$\sin\theta_4 \sin\theta_2 + \cos\theta_4 \cos\theta_2 = \frac{L_4{}^2 + L_2{}^2 + L_1{}^2 - L_3{}^2}{2L_4 L_2} + \frac{L_1}{L_2}\cos\theta_4 - \frac{L_1}{L_4}\cos\theta_2$$

$$\cos(\theta_4 - \theta_2) = \frac{L_4{}^2 + L_2{}^2 + L_1{}^2 - L_3{}^2}{2L_4 L_2} + \frac{L_1}{L_2}\cos\theta_4 - \frac{L_1}{L_4}\cos\theta_2$$

To simplify this equation, three dimensionless variables are defined. Let:

$$C_1 = \frac{L_4{}^2 + L_2{}^2 + L_1{}^2 - L_3{}^2}{2L_4 L_2}$$

$$C_2 = \frac{L_1}{L_2}$$

$$C_3 = \frac{L_1}{L_4}$$

Then Freudenstein's equation becomes:

$$C_1 + C_2 \cos\theta_4 - C_3 \cos\theta_2 = \cos(\theta_4 - \theta_2)$$

If θ_4 and θ_2 are known, then this equation is linear as far as C_1, C_2, and C_3 are concerned. If one of the link lengths is known, then the other three link lengths can be determined using C_1, C_2, and C_3. Two 4-bar linkages that are scaled versions of each other will have the same $\theta_4 - \theta_2$ relationship, thus these 4-bar linkage function generators are scale invariant. They are also rotation invariant so they may be oriented in any direction without affecting the $\theta_4 - \theta_2$ relationship.

Function generation using a 4-bar linkage consists of a relationship between $\Delta\theta_4$ and $\Delta\theta_2$ as seen in Figure 4.4. The starting orientation for the function generator starts at θ_{40} and θ_{20} and moves in the direction of $\Delta\theta_4$ and $\Delta\theta_2$. Let us assume that θ_{40} and θ_{20} are free choices and $\Delta\theta_4$ and $\Delta\theta_2$ are related through the specified function. For the initial design, let us assume that L_1 is a fixed length between the bearings A_o and B_o and that θ_1 is equal to 0° and along the x-axis. This leaves three unknowns, L_2, L_3, and L_4.

In applications where oscillating high torque needs to be transmitted between the input and the output, a 4-bar linkage is a better design than gears because it has a higher torque density ratio. **Torque density ratio** is defined as the ratio of the transmitted torque to the mechanism's

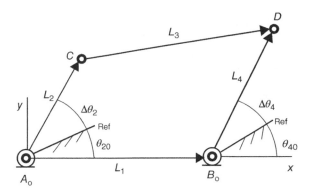

Figure 4.4 Relationship among angles

weight. Gears work well for transmitting continuous torque while linkages work better for transmitting oscillating torques.

4.4   Three-Point Matching Method for 4-Bar Linkage

The three-point matching method is a way to design a 4-bar linkage that goes through three specified points. The function's independent variable is initially selected using Chebyshev's spacing and is classified as free choices. The dependent variable is defined by the function. We will assume that $\Delta\theta_2$ is the independent variable and $\Delta\theta_4$ is the dependent variable.

The range for $\Delta\theta_4$ and $\Delta\theta_2$ are also free choices. Once these choices are made, Freudenstein's equation is used to create three linear equations with C_1, C_2, and C_3 as the unknowns. Many times the distance between the fixed bearings A_o and B_o is dictated so we will assume that L_1 is known, and then solve for the other three unknown lengths L_2, L_3, and L_4. If a link length comes out negative, it indicates that its actual angle is different from the estimated one in Figure 4.4 by 180°.

We need to realize that Chebyshev's spacing may not produce an optimum design, but it will be a good starting point for the design. After determining an initial design, more precision points can be added to the design, thus reducing the number of free choices by the same number.

Example 4.3   Replace a set of spur gears that transmits an oscillating torque with a 4-bar linkage

Problem: An 80-tooth and a 32-tooth gear set having a gear ratio of 2.5:1 is used to transmit an oscillating torque. The two fixed bearings are currently 28 in. apart since the spur gears have a diametric pitch of 2 teeth/in. The output ($\Delta\theta_4$) oscillates through 60° while the input ($\Delta\theta_2$) oscillates through 150°. The design engineer wants to replace these gears with a 4-bar linkage.

Solution:

1. Determine the values for the free choices in the 4-bar linkage design.

$$L_1 = 28 \text{ in. at } 0°, \ \theta_{20} = 35°, \text{ and } \theta_{40} = 135°.$$

2. Calculate the three precision points using Chebyshev spacing.

$$\theta_{2j} = \theta_{20} + \frac{\Delta\theta_2}{2}\left[1 - \cos\left(\frac{(2j-1)\cdot(90°)}{3}\right)\right], \quad j = 1,2,3$$

$$\theta_{21} = 35° + \frac{150}{2}\left[1 - \cos\left(\frac{(2-1)*90°}{3}\right)\right] = 35° + 10.0° = 45°$$

$$\theta_{22} = 35° + \frac{150}{2}\left[1 - \cos\left(\frac{(4-1)*90°}{3}\right)\right] = 35° + 75.0° = 110°$$

$$\theta_{23} = 35° + \frac{150}{2}\left[1 - \cos\left(\frac{(6-1)*90°}{3}\right)\right] = 35° + 140.0° = 175°$$

3. Using the specified function, calculate the dependent variable values.

$$\theta_{4j} = \theta_{40} + f(\theta_{2j} - \theta_{20}), \quad j = 1,2,3$$

$$\theta_{41} = 135° - \frac{10.0°}{2.5} = 131°$$

$$\theta_{42} = 135° - \frac{75.0°}{2.5} = 105°$$

$$\theta_{43} = 135° - \frac{140.0°}{2.5} = 79°$$

4. Use Freudenstein's equation to create three linear equations.

$$C_1 + C_2\cos\theta_4 - C_3\cos\theta_2 = \cos(\theta_4 - \theta_2)$$

$$\begin{bmatrix} 1 & \cos\theta_{41} & -\cos\theta_{21} \\ 1 & \cos\theta_{42} & -\cos\theta_{22} \\ 1 & \cos\theta_{43} & -\cos\theta_{23} \end{bmatrix} \begin{Bmatrix} C_1 \\ C_2 \\ C_3 \end{Bmatrix} = \begin{Bmatrix} \cos(\theta_{41} - \theta_{21}) \\ \cos(\theta_{42} - \theta_{22}) \\ \cos(\theta_{43} - \theta_{23}) \end{Bmatrix}$$

5. Insert the three precision data points into the three Freudenstein equations.

$$\begin{bmatrix} 1 & \cos(131°) & -\cos(45.0°) \\ 1 & \cos(105°) & -\cos(110°) \\ 1 & \cos(79.0°) & -\cos(175°) \end{bmatrix} \begin{Bmatrix} C_1 \\ C_2 \\ C_3 \end{Bmatrix} = \begin{Bmatrix} \cos(131° - 45.0°) \\ \cos(105° - 110°) \\ \cos(79.0° - 175°) \end{Bmatrix}$$

6. Solve the three equations for C_1, C_2, and C_3.

$$C_1 = -2.535, \ C_2 = -8.315, \text{ and } C_3 = 4.031$$

7. Knowing $L_1 = 28$ in., solve for link lengths L_2, L_3, and L_4.

$L_2 = -3.367$ in. or (3.367 in the opposite direction from guess)
$L_3 = 26.93$ in.
$L_4 = 6.947$ in.

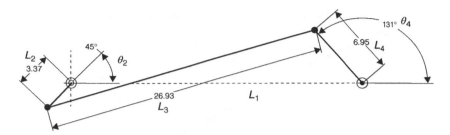

Figure 4.5 Designed 4-bar linkage

8. Assemble the 4-bar linkage (see Figure 4.5).
9. Analyze the 4-bar linkage to see how it behaves compared to the specified function.

θ_2 (deg)	θ_4 (deg)	θ_4 (deg)	% Error
35	133.2	135	−1.3
45	131	131	0.0
55	128.1	127	0.9
65	124.7	123	1.4
75	120.9	119	1.6
85	116.7	115	1.5
95	112.1	111	1.0
105	107.4	107	0.4
115	102.5	103	−0.5
125	97.6	99	−1.4
135	92.9	95	−2.2
145	88.5	91	−2.7
155	84.6	87	−2.8
165	81.3	83	−2.0
175	79	79	0.0
185	77.6	75	3.5

10. Plot the solution and compare it with the desired (see Figure 4.6). In this case the desired
    function is $\theta_4 = 149° - 0.400 \cdot \theta_2$

$$\text{The 4-bar linkage's function is } \theta_4 = 149.8° - 0.4088 \cdot \theta_2$$

    If this is not close enough to the desired function, then a redesign is in order.
11. Redesign if necessary by trying different values for the free choices.

Answers: $L_1 = 28''$, $L_2 = 3.367''$, $L_3 = 26.93''$, $L_4 = 6.947''$, $\theta_{20} = 35°$, and
$\theta_{40} = 135°$.

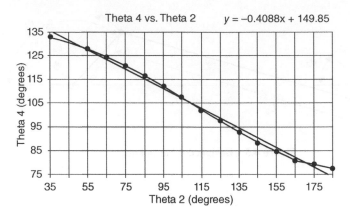

Figure 4.6   Plot of 4-bar design

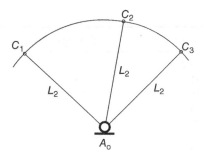

Figure 4.7   Locate bearing A_o

4.5   Design a 4-Bar Linkage for Body Guidance

This analytical approach will locate the pivot point or bearing given the endpoint of a specified link such as link 2. It assures that the pivot point is equidistant from the three specified points.

Consider three points, C_1, C_2, and C_3 in Figure 4.7. The distance between C_1 and A_o is given by the following equation.

$$L_2 = \sqrt{\left(C_{1x} - A_{ox}\right)^2 + \left(C_{1y} - A_{oy}\right)^2}$$

Likewise, the distance between C_2 and A_o and C_3 and A_o is:

$$L_2 = \sqrt{\left(C_{2x} - A_{ox}\right)^2 + \left(C_{2y} - A_{oy}\right)^2}$$

$$L_2 = \sqrt{\left(C_{3x} - A_{ox}\right)^2 + \left(C_{3y} - A_{oy}\right)^2}$$

If A_o is equidistance from each of the three points, then the three equations for L_2 must be equal.

$$\sqrt{(C_{1x}-A_{ox})^2 + (C_{1y}-A_{oy})^2} = \sqrt{(C_{2x}-A_{ox})^2 + (C_{2y}-A_{oy})^2}$$

$$\sqrt{(C_{2x}-A_{ox})^2 + (C_{2y}-A_{oy})^2} = \sqrt{(C_{3x}-A_{ox})^2 + (C_{3y}-A_{oy})^2}$$

Squaring both sides of these two equations, collecting like terms, and simplifying lead to:

$$(C_{1x}-A_{ox})^2 + (C_{1y}-A_{oy})^2 = (C_{2x}-A_{ox})^2 + (C_{2y}-A_{oy})^2$$

$$(C_{2x}-A_{ox})^2 + (C_{2y}-A_{oy})^2 = (C_{3x}-A_{ox})^2 + (C_{3y}-A_{oy})^2$$

then

$$C_{1x}{}^2 - 2C_{1x}A_{ox} + A_{ox}{}^2 + C_{1y}{}^2 - 2C_{1y}A_{oy} + A_{oy}{}^2 = C_{2x}{}^2 - 2C_{2x}A_{ox} + A_{ox}{}^2 + C_{2y}{}^2 - 2C_{2y}A_{oy} + A_{oy}{}^2$$

$$C_{2x}{}^2 - 2C_{2x}A_{ox} + A_{ox}{}^2 + C_{2y}{}^2 - 2C_{2y}A_{oy} + A_{oy}{}^2 = C_{3x}{}^2 - 2C_{3x}A_{ox} + A_{ox}{}^2 + C_{3y}{}^2 - 2C_{3y}A_{oy} + A_{oy}{}^2$$

simplifying

$$2(C_{2x}-C_{1x})A_{ox} + 2(C_{2y}-C_{1y})A_{oy} = (C_{2x}{}^2 + C_{2y}{}^2 - C_{1x}{}^2 - C_{1y}{}^2)$$

$$2(C_{3x}-C_{2x})A_{ox} + 2(C_{3y}-C_{2y})A_{oy} = (C_{3x}{}^2 + C_{3y}{}^2 - C_{2x}{}^2 - C_{2y}{}^2)$$

These two equations are linear with respect to A_{ox} and A_{oy} so Cramer's Rule or substitution can be used to solve for the (x, y) location of bearing A_o. In a similar fashion, if the other end of link 3 is known such as D_1, D_2, and D_3, then the (x, y) location of B_o can be solved for as follows.

$$2(D_{2x}-D_{1x})B_{ox} + 2(D_{2y}-D_{1y})B_{oy} = (D_{2x}{}^2 + D_{2y}{}^2 - D_{1x}{}^2 - D_{1y}{}^2)$$

$$2(D_{3x}-D_{2x})B_{ox} + 2(D_{3y}-D_{2y})B_{oy} = (D_{3x}{}^2 + D_{3y}{}^2 - D_{2x}{}^2 - D_{2y}{}^2)$$

Example 4.4   Determine the proper location for bearings A_o and B_o given the coupler link in three different locations

Problem: In designing a pressure-sealing door over an eight-foot opening, it is important that the door lines up with the edges of the opening and then moves forward to seal the opening. Design a 4-bar linkage that will move the door away from the sealed entrance and then rotate the door approximately 90°. In position 1, $C_1 = (3.0, 5.4)$ ft and $D_1 = (3.0, 13.0)$ ft. In position 2, $C_2 = (5.0, 2.7)$ ft and $D_2 = (12.6, 2.7)$ ft, and in position 3, $C_3 = (5.0, 0.0)$ ft and $D_3 = (12.6, 0.0)$ ft (see Figure 4.8).

Solution:

1. Locate the sealing door in its three positions and determine where of the door you want to attach the 4-bar linkage.
2. Substitute values into the following two equations, then solve for A_{ox} and A_{oy}.

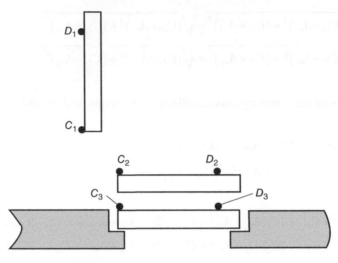

Figure 4.8   Sealing a large 8-foot door

$$2(C_{2x}-C_{1x})A_{ox} +2(C_{2y}-C_{1y})A_{oy} = (C_{2x}{}^2+C_{2y}{}^2-C_{1x}{}^2-C_{1y}{}^2)$$

$$2(C_{3x}-C_{2x})A_{ox} +2(C_{3y}-C_{2y})A_{oy} = (C_{3x}{}^2+C_{3y}{}^2-C_{2x}{}^2-C_{2y}{}^2)$$

$$2(5.0-3.0)A_{ox} +2(2.7-5.4)A_{oy} = (5.0^2+2.7^2-3.0^2-5.4^2)$$

$$2(5.0-5.0)A_{ox} +2(0.0-2.7)A_{oy} = (5.0^2+0.0^2-5.0^2-2.7^2)$$

solving

$$A_{ox} =0.355 \text{ ft}$$

$$A_{oy} = 1.35 \text{ ft}$$

3. Substitute values into the following two equations, then solve for B_{ox} and B_{oy}.

$$2(D_{2x}-D_{1x})B_{ox} +2(D_{2y}-D_{1y})B_{oy} = (D_{2x}{}^2+D_{2y}{}^2-D_{1x}{}^2-D_{1y}{}^2)$$

$$2(D_{3x}-D_{2x})B_{ox} +2(D_{3y}-D_{2y})B_{oy} = (D_{3x}{}^2+D_{3y}{}^2-D_{2x}{}^2-D_{2y}{}^2)$$

$$2(12.6-3.0)B_{ox} +2(2.7-13.0)B_{oy} = (12.6^2+2.7^2-3.0^2-13.0^2)$$

$$2(12.6-12.6)B_{ox} +2(0.0-2.7)B_{oy} = (12.6^2+0.0^2-12.6^2-2.7^2)$$

solving

$$B_{ox} =0.826 \text{ ft}$$

$$B_{oy} = 1.35 \text{ ft}$$

4. Determine the size of links 2 and 4, which connect between the door pivots and the fixed bearings.

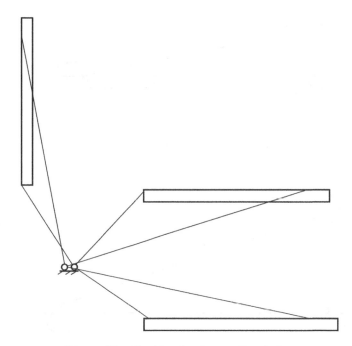

Figure 4.9   Checking the door sealing design

$$L_2 = \sqrt{(C_{1x}-A_{ox})^2 + (C_{1y}-A_{oy})^2} = \sqrt{(3.0-0.355)^2 + (5.4-1.35)^2} = 4.837\,\text{ft}$$

$$L_4 = \sqrt{(D_{1x}-B_{ox})^2 + (D_{1y}-B_{oy})^2} = \sqrt{(3.0-0.826)^2 + (13.0-1.35)^2} = 11.85\,\text{ft}$$

Answers: $\boxed{A_o = (0.355,1.35)\,\text{ft}}$, $\boxed{B_o = (0.826,1.35)\,\text{ft}}$, $\boxed{L_2 = 4.837\,\text{ft}}$, and $\boxed{L_4 = 11.85\,\text{ft}}$.

The resulting linkage is shown in Figure 4.9 in its three configurations. Although the linkage can be assembled in its three positions, it is not possible to move between position two and position three without taking the linkage apart and then reassembling it in the closed position. For this reason, this linkage design is not a workable solution. This problem is known as "branch defect" since the solution lies in two different branches of the linkage. If the door was tilted slightly in the second position, a workable solution may be found. Remember, not all solutions are workable solutions. You must check each solution you find as a design engineer.

4.6   Function Generation for Slider-Crank Mechanisms

In the case where you want the input variable to move along a rotational path while the output variable moves along a straight-line path, a slider-crank mechanism is your best choice. Assume $y = f(\Delta\theta_2)$ (see Figure 4.10).

We will use Freudenstein's equation to relate the output variable "y" to the input variable "$\Delta\theta_2$." Figure 4.11 shows a slider-crank mechanism and its vector loop.

The vector loop equation follows along with the y-component and x-component equations.

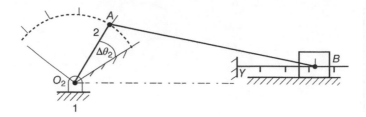

Figure 4.10 Slider-crank function generator

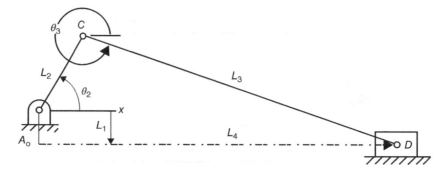

Figure 4.11 Slider-crank mechanism and its vector loop

$$L_2 + L_3 - L_4 - L_1 = 0$$

$$L_2 \sin\theta_2 + L_3 \sin\theta_3 + L_1 = 0$$

$$L_2 \cos\theta_2 + L_3 \cos\theta_3 - L_4 = 0$$

If we place the L_3 terms on the left side of each equation, then square each equation and add them together, we get:

$$L_3 \sin\theta_3 = -L_1 - L_2 \sin\theta_2$$

$$L_3 \cos\theta_3 = L_4 - L_2 \cos\theta_2$$

then

$$L_3{}^2 \sin^2\theta_3 = L_1{}^2 + 2L_1 L_2 \sin\theta_2 + L_2{}^2 \sin^2\theta_2$$

$$L_3{}^2 \cos^2\theta_3 = L_4{}^2 - 2L_4 L_2 \cos\theta_2 + L_2{}^2 \cos^2\theta_2$$

adding

$$L_3{}^2 = L_1{}^2 + L_4{}^2 + 2L_1 L_2 \sin\theta_2 - 2L_4 L_2 \cos\theta_2 + L_2{}^2$$

or

$$\left(L_3{}^2 - L_1{}^2 - L_2{}^2\right) - (2L_1 L_2)\sin\theta_2 + (2L_2)L_4 \cos\theta_2 = L_4{}^2$$

Let three constants be equal to anything that is not θ_2 or L_4 since these two variables are the input and output of the function generator.

$$C_1 = \left(L_3{}^2 - L_1{}^2 - L_2{}^2\right)$$

$$C_2 = \left(-2L_1L_2\right)$$

$$C_3 = \left(2L_2\right)$$

then

$$C_1 + C_2 \sin\theta_2 + C_3 L_4 \cos\theta_2 = L_4{}^2$$

Unlike the 4-bar linkage equation, function generation for a slider-crank mechanism is not scale invariant because C_1, C_2, and C_3 are not dimensionless.

4.7   Three-Point Matching Method for Slider-Crank Mechanism

The point-matching method with the above equation is a method by which a slider-crank mechanism can be designed to match three precision points, thus approximating a function generator. The input can be the rotatory motion of link 2, the crank, or the linear motion of link 4, the slider. The three precision points can be determined by Chebyshev spacing or at random since they are free choices. The starting angle of link 2 and the starting position of the slider are also free choices. This leaves plenty of room for optimizing your design. The following set of three linear equations will be used to solve for the three unknowns C_1, C_2, and C_3, and then back substitution will be used to determine the unknown link lengths, L_1, L_2, and L_3.

$$\begin{vmatrix} 1 & \sin\theta_{21} & L_{41}\cos\theta_{21} \\ 1 & \sin\theta_{22} & L_{42}\cos\theta_{22} \\ 1 & \sin\theta_{23} & L_{43}\cos\theta_{23} \end{vmatrix} \begin{Bmatrix} C_1 \\ C_2 \\ C_3 \end{Bmatrix} = \begin{Bmatrix} L_{41}{}^2 \\ L_{42}{}^2 \\ L_{43}{}^2 \end{Bmatrix}$$

solve for $C_1, C_2,$ and C_3, then back substitute

$$L_2 = \frac{C_3}{2}$$

$$L_1 = \frac{-C_2}{2L_2}$$

$$L_3 = \sqrt{\left(C_1 + L_1{}^2 + L_2{}^2\right)}$$

The following example illustrates this procedure.

Example 4.5   Design a slider-crank function generator with a rotational input and a linear output motion

Problem: Design a slider-crank function generator for the following function, $y = 1.5x^2$ for x ranging from 0.5 to 1.5 inclusive. Variable "x" is positive counterclockwise and variable "y" is positive right to left.

Solution:

1. Calculate the range of the input and output variables.

   $x_{min} = 0.5$, $y = 0.375$ and $x_{max} = 1.5$, $y = 3.375$, thus $(\Delta x = 1.000)$ and $(\Delta L_4 = -3.000)$.

2. Determine three precision points using Chebyshev spacing for the input variable, x.

$$x_j = x_{min} + \frac{x_{max} - x_{min}}{2}\left[1 - \cos\left(\frac{(2j-1)\cdot(90°)}{n}\right)\right], \quad j = 1, 2, \ldots, n$$

$$x_1 = 0.567$$
$$x_2 = 1.000$$
$$x_3 = 1.433$$

3. Determine the three precision points for the output variable, y.

$$y_i = 1.5(x_i)^2$$
$$y_1 = 1.5\cdot(0.567)^2 = 0.482$$
$$y_2 = 1.5\cdot(1.000)^2 = 1.500$$
$$y_3 = 1.5\cdot(1.433)^2 = 3.080$$

4. Determine the desired range of motion for the input and output variables.

   Let $\Delta\theta_2 = 60°$ and $\Delta y = -3.000$ in.

5. Choose the initial angle θ_{20} and the initial distance the slider is away from the pivot, L_{40}.

   $\theta_{20} = 55°$ and $L_{40} = 6.00$ in. (see Figure 4.12).

6. Determine the three precision input points to be used in designing the slider-crank function generator.

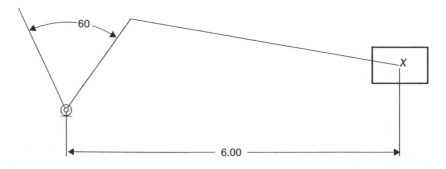

Figure 4.12   Initial layout of slider-crank mechanism

$$\theta_{2i} = \theta_{20} + \frac{x_i - x_{min}}{\Delta x}(\Delta \theta_2) = 55° + \frac{x_i - 0.5}{1.000}(60°)$$

$$\theta_{21} = 59°$$

$$\theta_{22} = 85°$$

$$\theta_{23} = 111°$$

7. Determine the three precision output points. Letting (ΔL_4) be a negative 3.000 in.

$$L_{4i} = L_{40} + \frac{y_i - y_{min}}{\Delta y}(\Delta L_4) = 6.000'' + \frac{y_i - 0.375}{3.000}(-3.000'')$$

$$L_{41} = 5.893''$$

$$L_{42} = 4.875''$$

$$L_{43} = 3.295''$$

8. Create the three linear equations to be solved.

$$\begin{vmatrix} 1 & \sin\theta_{21} & L_{41}\cos\theta_{21} \\ 1 & \sin\theta_{22} & L_{42}\cos\theta_{22} \\ 1 & \sin\theta_{23} & L_{43}\cos\theta_{23} \end{vmatrix} \begin{Bmatrix} C_1 \\ C_2 \\ C_3 \end{Bmatrix} = \begin{Bmatrix} L_{41}^2 \\ L_{42}^2 \\ L_{43}^2 \end{Bmatrix}$$

9. Solve for the three unknowns.

$$C_1 = -20.51$$

$$C_2 = 41.71$$

$$C_3 = 6.422$$

10. Using back substitution, solve for the three lengths.

$L_2 = 3.211$ in.

$L_1 = -6.495$ in. (positive is down, see Figure 4.11)

$L_3 = 5.655$ in.

11. Check the results at the three precision points.

$$L_4^2 - (C_3\cos\theta_2)L_4 - (C_1 + C_2\sin\theta_2) = 0$$

thus

$$L_4 = \frac{(C_3\cos\theta_2) + \sqrt{(C_3\cos\theta_2)^2 + 4(C_1 + C_2\sin\theta_2)}}{2}$$

12. Check the results throughout the range and redesign if necessary (see Figure 4.13).

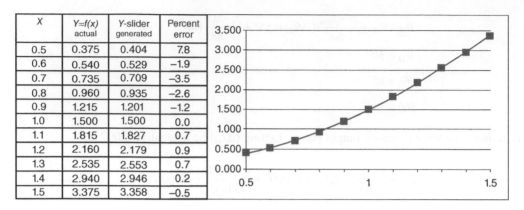

X	Y=f(x) actual	Y-slider generated	Percent error
0.5	0.375	0.404	7.8
0.6	0.540	0.529	−1.9
0.7	0.735	0.709	−3.5
0.8	0.960	0.935	−2.6
0.9	1.215	1.201	−1.2
1.0	1.500	1.500	0.0
1.1	1.815	1.827	0.7
1.2	2.160	2.179	0.9
1.3	2.535	2.553	0.7
1.4	2.940	2.946	0.2
1.5	3.375	3.358	−0.5

Figure 4.13   Slider-crank function generator check

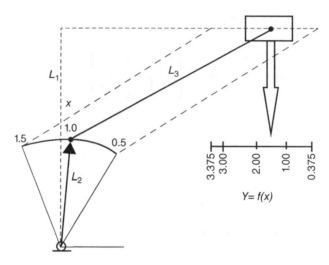

Figure 4.14   Slider-crank function generator

Although the slider-crank generated function is slightly different from the actual function, the two curves lie very close to each other on the graph except for the first point which differs by almost 8%. I would consider this an acceptable design solution (see Figure 4.14).

Solution: $\boxed{\text{Crank} = L_2 = 3.211''}$, $\boxed{\text{connector} = L_3 = 5.655''}$, $\boxed{\text{offset} = L_1 = -6.495''.}$
$\boxed{X_{\min} = 0.5 \text{ at } 55°}$, $\boxed{X_{\max} = 1.5 \text{ at } 115°}$,
$\boxed{3.00'' \text{ linear } Y\text{-scale is labeled from } 0.375 \text{ to } 3.375}$.

Problems

4.1   Determine the Chebyshev spacing for the variable, x, that goes from -1 to $+5$ using three precision points.

4.2   Determine the Chebyshev spacing for the variable, x, that goes from 1 to 6 using four precision points.

4.3   Determine the Chebyshev spacing for the variable, θ, that goes from 15 to $85°$ using three precision points.

4.4   For the function $y = 2x^{1.8}$ with x ranging from 1.5 to 2.5, determine the Chebyshev spacing for both variables, x and y, using five precision points.

4.5   For the function $y = 2x^{1.8}$ with x ranging from 1.5 to 2.5, determine the Chebyshev spacing for both variables using three precision points. Assume that the independent variable, θ_2, ranges from 210 to $270°$, while the dependent variable, θ_4, ranges from 235 to $305°$ relative to the x-axis.

4.6   Design a 4-bar linkage function generator according to the data given in Problem 4.5 above. Assume the distance between the fixed pivots is 4.00 in. and they are located on the same horizontal line.

4.7   Design a 4-bar linkage function generator that converts the input angle between 0 and $90°$ to the sinc of the angle with a range of 0.0–1.0. The input must rotate through $90°$ while the output is a linear scale rotating through $60°$. Assume the distance between the fixed pivots is 4.00 in. and they are located on the same horizontal line. Let the input link 2 vary between 45 and $135°$ while output link 4 varies between 130 and $190°$ relative to the x-axis.

4.8   Design a 4-bar linkage function generator that converts the input angle $20–70°$ to the tangent of the angle. The input must rotate through $50°$ while the output is a linear scale rotating through $90°$. Assume the distance between the fixed pivots is 4.00 in. and they are located on the same horizontal line. Let the input link 2 vary between 20 and $70°$ while output link 4 varies between 45 and $135°$ relative to the x-axis.

4.9   Design a 4-bar linkage that will replace two oscillating spur gears with 40 teeth and 72 teeth. The center distance between the gears was 7.00 in. The smaller gear rotated through $72°$. Let the input link 2 rotate counterclockwise between -10 and $62°$ while the output link 4 moves clockwise between 270 and $230°$.

4.10 Design a 4-bar linkage whose coupler link goes through the three positions specified. The coupler link is 100 mm long with endpoints labeled C and D. $C_1 = (20, 20)$ mm with $\theta_{31} = 20°$. $C_2 = (35, 40)$ mm with $\theta_{32} = 40°$. $C_3 = (48, 60)$ mm with $\theta_{32} = 70°$. What are the sizes of links 2 and 4? Where are the fixed pivots located?

4.11 Design a 4-bar linkage whose coupler link goes through the three positions specified. The coupler link is 5.00 in. long with endpoints labeled C and D. $C_1 = (2.00, 2.00)$ in. with $\theta_{31} = 25°$. $C_2 = (2.50, 3.00)$ in. with $\theta_{32} = 45°$. $C_3 = (3.50, 4.00)$ in. with $\theta_{33} = 75°$. What are the sizes of links 2 and 4? Where are the fixed pivots located?

4.12 Design an Ackerman steering 4-bar linkage using the point matching method. Allow the vehicle to turn left or right as it goes around a curve while keeping the front tires perpendicular to the radius of curvature as shown in Figure 4.15. Assume the distance between the centers of the front tires, W, is 60 in., and the wheelbase, L, is 116 in. Since driving straight is most common and critical, select one precision point with the wheel pointed straight ahead $(\theta_2 = \theta_4 = 0°)$ at $R =$ infinity. When turning left, let $R = 25$ ft. When turning right, let $R = -25$ ft. Let link $L_1 = 48$ in. with links 2 and 4 being $67°$ from L_1 as shown below. The relationship for the Ackerman steering is:

$$\frac{L}{\tan\theta_2} + W = \frac{L}{\tan\theta_4}$$

where,

$$\theta_2 = \tan^{-1}\left(\frac{L}{R - \frac{W}{2}}\right)$$

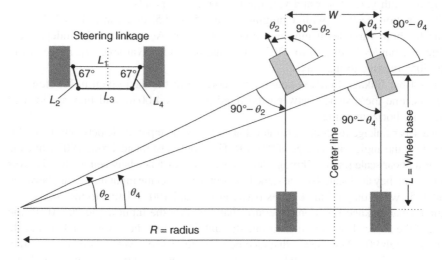

Figure 4.15 Ackerman steering linkage

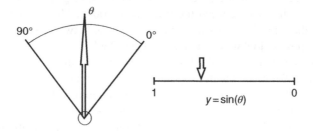

Figure 4.16 Problem 4.13

4.13 Design a slider-crank function generator that converts the input angle, θ, from 0 to 90° to the sine of the angle along a straight-line scale marked 1–0, left to right. The output scale should be 4.00 in. with its farthest point 6.00 in. right of the fixed bearing. The crank angle should start at 45° and end at 135° (see Figure 4.16).

4.14 Design a slider-crank function generator that converts the input angle, θ, 20–70° to the tangent of the angle along a straight-line scale marked 3.0–0, right to left. The output scale should be 75 mm long with its farthest point 150 mm right of the fixed bearing. The crank angle should start at −25° and end at 75° relative to the x-axis. Function generator design should be similar to Figure 4.16 above.

Further Reading

Hall, Allen S., Jr., *"Kinematics and Linkage Design"*, Prentice-Hall, Inc., Englewood Cliffs, NJ, 1961.
Mabie, Hamilton H., and Reinholtz, Charles F, *"Mechanisms and Dynamics of Machinery"*, John Wiley & Sons, New York, 1987.

5

Velocity Analysis

5.1 Introduction

Velocity is important because it affects the time required to perform a given operation, such as the machining of a part. Power is the product of force and linear velocity or torque and angular velocity. For the transmission of a given amount of power, the forces and stresses in the various links of a mechanism can be reduced by altering the velocities through a change in the size of the links. Friction and wear on machine parts also depend upon velocity. Determination of the velocities of a mechanism is required if an acceleration analysis is to be performed.

A straight line through an object and its axis of rotation can define an object. Any point not located at the axis of rotation will travel in a circular path around the axis of rotation (see Figure 5.1). The relative linear velocity of this point will be the product of the angular velocity of the object times the distance the point is away from the axis of rotation. The direction of this relative linear velocity will always be perpendicular to the line between the axis of rotation and the point and in the direction of the angular velocity. Note that linear velocity must be represented by a vector containing both a magnitude and a direction:

$$\overline{V_{A/O}} = \omega \cdot d_{OA}(\perp d_{OA})$$

$$\overline{V_{B/O}} = \omega \cdot d_{OB}(\perp d_{OB})$$

The equations above do not preclude the fact that point O may be moving since they are relative linear velocities. If the velocity of point "O" is not zero, then the linear velocities of points "A" and "B" relative to the ground become

$$\overline{V_{A/\text{ground}}} = \omega \cdot d_{OA}(\perp d_{OA}) + \overline{V_{O/\text{ground}}}$$

$$\overline{V_{B/\text{ground}}} = \omega \cdot d_{OB}(\perp d_{OB}) + \overline{V_{O/\text{ground}}}$$

Design and Analysis of Mechanisms: A Planar Approach, First Edition. Michael J. Rider.
© 2015 John Wiley & Sons, Ltd. Published 2015 by John Wiley & Sons, Ltd.

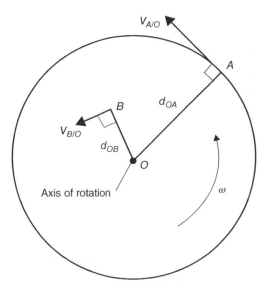

Figure 5.1 Rotation about an axis

We can apply this concept to the velocity analysis of planar linkages. Before we do this, one other concept needs to be discussed. An instant center is a point on two different objects that have exactly the same linear velocity vector. That is a point on two different objects that have the same velocity magnitude and direction. If two links were pinned together, then the location of the pin would be an instant center for these two objects since the pin is located in both objects and must travel exactly the same way through space.

5.2   Relative Velocity Method

The relative velocity method uses the concept that every point on an object can be defined if the axis of rotation and a line from this axis to any point are defined. Just as with relative position vectors as described in Chapter 2, relative velocity vectors can be treated the same way.

The slider–slider linkage (Figure 5.2) will be used to illustrate this concept. Since the linkage has been built or drawn previously, then the length of link 3, the connector link, and its angle of orientation relative to the +x-axis should be known. Let us assume that the linear velocity vector of block 2 containing point A is to the right and known. The velocity vector of block 4 containing point B must be vertical. The unknowns are the linear velocity vector of block 4 and the angular velocity vector of link 3 containing points A and B. Note that point A is an instant center for objects 2 and 3, while point B is an instant center for objects 3 and 4:

$$\overline{V_B} = \overline{V_A} + \overline{V_{B/A}}$$

$$V_B \angle 270° = V_A \angle 0° + \omega_3 \cdot L_{AB} \angle (\theta_3 + 90°)$$

This vector equation can be solved graphically since it is two-dimensional or it can be solved analytically. To solve this problem graphically, you will need a protractor and a scale marked

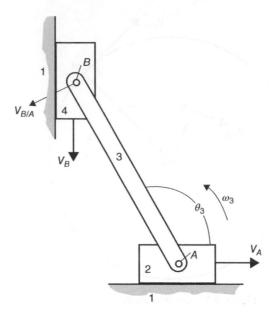

Figure 5.2   Slider–slider linkage

off in inches or millimeters. Starting at the origin and using the left side of the equation, draw any known part of this term. In this case, the angle of the resulting vector is known.

Determine the known quantities on the right side of the equation. In this case, the linear velocity of point A and its direction are known along with the direction of the relative velocity vector, $V_{B/A}$. Draw the known vector starting at the origin, and then add to its tip the direction of the relative velocity vector. If the magnitude of the known velocity vector is very large or very small, a scale factor may be used so the vectors fit in the working space of the page. The intersection of the two directional lines represents the solution to this problem. Measure the length of these lines, and then determine the answers using the applied scale factor.

Example 5.1   Velocity analysis of a slider–slider linkage using the relative velocity method

Problem: Block 2 is traveling at 4.00 in./s to the right (see Figure 5.2). Link 3 is 5.38 in. long from A to B and oriented at 120° from the x-axis. Determine the linear velocity vector of block 4 and the angular velocity vector of link 3.

Solution: Write the relative velocity equation:

$$\overline{V_B} = \overline{V_A} + \overline{V_{B/A}}$$

$$V_B \angle 270° = 4.00 \angle 0° + 5.38\omega_3 \angle (120° + 90°)$$

$$V_B \angle 270° = 4.00 \angle 0° + 5.38\omega_3 \angle 210°$$

Draw the known vector quantities (either magnitude or direction) for each side of the equation starting at the origin each time (see Figure 5.3). The intersection of the two dashed lines is the solution. Measure the lengths of these two lines, and then relate them to the unknowns:

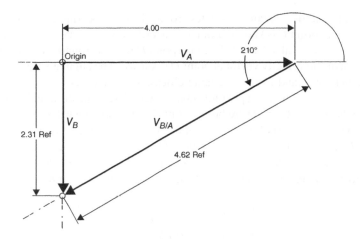

Figure 5.3   Graphical solution

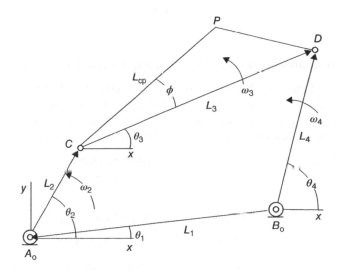

Figure 5.4   Four-bar linkage

$$V_B = 2.31 \, \text{in./s} \downarrow$$

$$V_{B/A} = 4.62 \, \text{in./s} = 5.38\omega_3$$

$$\omega_3 = 4.62/5.38 = 0.859 \, \text{rad/s c.c.w.}$$

Answer: $\boxed{\overline{V_B} = 2.31 \, \text{in} \cdot /\text{s} \downarrow}$ and $\boxed{\overline{\omega_3} = 4.62/5.38 = 0.859 \, \text{rad/s c.c.w.}}$

Next, consider the 4-bar linkage shown in Figure 5.4. Fixed bearings are located at A_o and B_o so the velocity of these two points is zero. The input for this 4-bar linkage is link 2; thus, we know the angular velocity vector, ω_2. Point C rotates around the fixed bearing at A_o so the direction of the velocity vector at point C is known. Point C is also the instant center for links 2 and 3. Point D is the instant center for links 3 and 4. Because point D rotates around the fixed bearing B_o, the general direction of the velocity vector at point D is known. The angular velocity vectors for links 3 and 4 are unknown and assumed counterclockwise (or c.c.w.).

For a given angular position of link 2, a position analysis will determine the angular positions of links 3 and 4. Note that points C and D are both on link 3 so the relative velocity method can be used to relate the linear velocities of these two points. The linear velocity vector for point D is equal to the length L_4 times ω_4 at an angle of $\theta_4 + 90°$. The linear velocity vector for point C is equal to the length L_2 times ω_2 at an angle of $\theta_2 + 90°$. The linear relative velocity vector for point D relative to point C is equal to the length L_3 times ω_3 at an angle of $\theta_3 + 90°$:

$$\overline{V_D} = \overline{V_C} + \overline{V_{D/C}}$$

$$\overline{V_D} = L_4 \cdot \omega_4 \angle (\theta_4 + 90°)$$

$$\overline{V_C} = L_2 \cdot \omega_2 \angle (\theta_2 + 90°)$$

$$\overline{V_{D/C}} = L_3 \cdot \omega_3 \angle (\theta_3 + 90°)$$

Example 5.2   Velocity analysis of a 4-bar linkage using the relative velocity method

Problem: A 4-bar linkage (Figure 5.4) has link lengths of $L_1 = 3.00$ in., $L_2 = 1.25$ in., $L_3 = 3.25$ in., and $L_4 = 2.00$ in. Bearing B_o is located at a $10°$ angle from bearing A_o. Using the current position, $\theta_2 = 60°$, the position analysis has determined that $\theta_3 = 24.3°$ and $\theta_4 = 71.5°$. Currently, link 2 is traveling at 5.50 rad/s counterclockwise. What are the angular velocity vectors for links 3 and 4?

Solution: Write the relative velocity equation and its supporting terms:

$$\overline{V_D} = \overline{V_C} + \overline{V_{D/C}}$$

$$\overline{V_D} = L_4 \cdot \omega_4 \angle (\theta_4 + 90°) = 2.00 \cdot \omega_4 \angle (71.5° + 90°) = 2.00 \cdot \omega_4 \angle 161.5°$$

$$\overline{V_C} = L_2 \cdot \omega_2 \angle (\theta_2 + 90°) = 1.25 * 5.50 \angle (60° + 90°) = 6.875 \angle 150°$$

$$\overline{V_{D/C}} = L_3 \cdot \omega_3 \angle (\theta_3 + 90°) = 3.25 \cdot \omega_3 \angle (24.3° + 90°) = 3.25 \cdot \omega_3 \angle 114.3°$$

Draw the known vector quantities (either magnitude or direction) for each side of the equation starting at the origin each time (see Figure 5.5). The intersection of the two dashed lines is the solution. Note that the intersection point for the $V_{D/C}$ line is in the opposite direction; thus, the angular velocity vector for link 3 will be clockwise (or c.w.) instead of counterclockwise. Measure the lengths of these two lines, and then relate them to the unknowns:

$$V_D = 5.47 \text{ in./s} = 2.00 \omega_4$$

$$\omega_4 = 5.47/2.00 = 2.73 \text{ rad/s c.c.w.}$$

$$V_{D/C} = -1.87 \text{ in./s} = 3.25 \omega_3$$

$$\omega_3 = -1.87/3.25 = 0.575 \text{ rad/s c.w.}$$

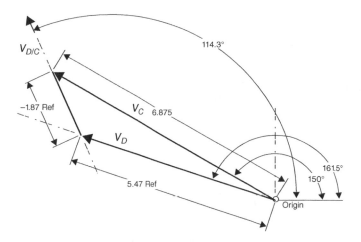

Figure 5.5   Graphical solution

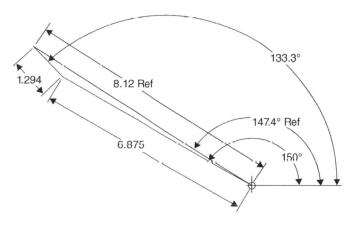

Figure 5.6   Velocity solution for coupler point

Answer: $\boxed{\overline{\omega_3} = 0.575\,\text{rad/s c.w.}}$ and $\boxed{\overline{\omega_4} = 2.73\,\text{rad/s c.c.w.}}$

If the linear velocity vector for point P (Figure 5.4) is required, then the relative velocity method can be used to determine it as follows. Figure 5.6 represents the graphical solution:

$$\overline{V_P} = \overline{V_C} + \overline{V_{P/C}}$$

$$\overline{V_P} = L_2 \cdot \omega_2 \angle (\theta_2 + 90°) + L_{CP} \cdot \omega_3 \angle (\theta_3 + \phi + 90°)$$

$$\overline{V_P} = 1.25 * 5.50 \angle (60° + 90°) + 2.25 \cdot 0.575 \angle (24.3° + 19.0° + 90°)$$

$$\overline{V_P} = 6.875 \angle 150° + 1.294 \angle 133.3° = 8.12\,\text{in./s} \angle 147.4°$$

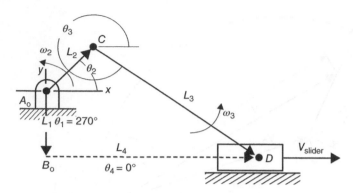

Figure 5.7   Slider–crank mechanism

Before moving on to the next velocity method, let us look at a slider–crank mechanism and how it can be solved using the relative velocity method (see Figure 5.7). Note that the angular velocity vector was assumed counterclockwise and the linear velocity vector of the slider was assumed to the right. Common sense indicates that the linear velocity vector for the slider is to the left for this example, but the assumption is still valid.

Once the slider–crank mechanism has been defined, a position analysis can be used to determine the angular position of link 3 (θ_3) and the linear position of the slider (L_4). Once again, link 2, the crank, is assumed to be the input link; thus, the angular velocity vector for link 2 is known. The linear velocity vector for point C can be determined since it travels in a circle around the fixed bearing A_o. The unknowns are the angular velocity vector of link 3 and the linear velocity vector of the slider. Points C and D are both on link 3 so the relative velocity vector method can be applied to these two points:

$$\overline{V_D} = \overline{V_C} + \overline{V_{D/C}}$$

$$\overline{V_D} = V_D \angle 0°$$

$$\overline{V_C} = L_2 \cdot \omega_2 \angle(\theta_2 + 90°)$$

$$\overline{V_{D/C}} = L_3 \cdot \omega_3 \angle(\theta_3 + 90°)$$

Starting at the origin, draw a horizontal line that will represent the velocity of the slider, V_D. Starting at the origin again, draw a vector that represents the velocity of point C. Both its magnitude and its direction are known. At the tip of this vector, draw a line that represents the direction for the relative velocity vector, $V_{D/C}$. The intersection of these two lines represents the solution. If the intersection is in the opposite direction from the assumed direction, then the original assumptions for the velocity directions were incorrect.

Example 5.3   Velocity analysis of a slider–crank mechanism using the relative velocity method

Problem: A slider–crank linkage (Figure 5.7) has link lengths of $L_1 = 2.10$ in., $L_2 = 2.00$ in., and $L_3 = 6.50$ in. Bearing A_o is located at the origin. For the current position of $\theta_2 = 43°$, the angle $\theta_3 = 327.8°$ and the length $L_4 = 6.963$ in. Note $\theta_1 = 270°$ and $\theta_4 = 0°$. Link 2 is currently traveling at 7.00 rad/s counter clockwise. Determine the angular velocity vector for link 3 (ω_3) and the linear velocity vector for the slider (V_{slider}).

Solution: Write the relative velocity equation and its supporting terms:

$$\overline{V_D} = \overline{V_C} + \overline{V_{D/C}}$$

$$V_D\angle 0° = L_2\omega_2\angle(\theta_2 + 90°) + L_3\omega_3\angle(\theta_3 + 90°)$$

$$V_D\angle 0° = 2.00*7.00\angle(43° + 90°) + 6.50\omega_3\angle(327.8° + 90°)$$

$$V_D\angle 0° = 14.00\angle 133° + 6.50\omega_3\angle 57.8°$$

Draw the known vector quantities (either magnitude or direction) for each side of the equation starting at the origin each time. The intersection of the two dashed lines is the solution. Measure the lengths of these two lines, and then relate them to the unknowns (see Figure 5.8):

$$V_D = -16.0\,\text{in./s} \rightarrow \text{or } 16.0\,\text{in./s} \leftarrow$$

$$V_{D/C} = -12.10\,\text{in./s} = 6.50\omega_3$$

$$\omega_3 = -12.10/6.50 = 1.86\,\text{rad/s c.w.}$$

Answer: $\boxed{\overline{V_D} = 16.0\,\text{in./s} \leftarrow}$ and $\boxed{\overline{\omega_3} = 1.86\,\text{rad/s c.w.}}$

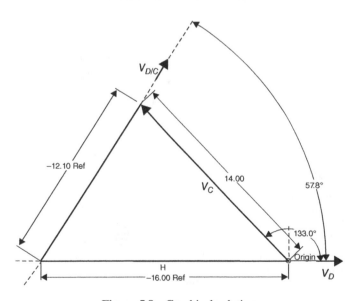

Figure 5.8   Graphical solution

5.3 Instant Center Method

Typically, the velocities of a mechanism are determined at numerous positions of the links. A link having plane motion is considered as rotating at that instant about some point in the plane of motion. This point is the center of rotation for the link and may or may not lie within the link itself. Some links may have their center of rotation as a stationary point, while for others the center of rotation moves. The term instant center of zero velocity is used to denote the center of rotation of the body about a stationery point at this instant.

An instant center is a point common to two objects having the same linear velocity in both magnitude and direction. If one of the bodies is fixed, then that instant center is called an instant center of zero velocity. Both of these definitions are important since we will make use of them in locating instant centers for various mechanisms.

For the 4-bar linkage shown in Figure 5.9, each pin connection is an instant center. It is customary to designate these centers using the numbers of the links that are pinned together at these points. Thus, the point on link 1 which link 2 rotates about is labeled IC_{12} (also labeled A_o in Figure 5.9). IC_{12} remains fixed since link 1 is considered the ground and does not move. This instant center is referred to as a fixed instant center. Point C is IC_{23} because it is a point on link 2 that link 3 rotates about or a point on link 3 that link 2 rotates about. Instant centers IC_{23} and IC_{34} are called moving instant centers since they move relative to ground.

For the slider–slider linkage shown in Figure 5.10, the pin connections at points A and B are instant centers IC_{23} and IC_{34}. All points on object 2 move along a straight horizontal line; thus, lines perpendicular to their velocities will be vertical parallel lines. Recalling that parallel lines intersect at infinity, it follows that IC_{12} is located at infinity above or below object 2. Similarly, IC_{14} is located on a horizontal line to the left or right of object 4 at infinity. In summary, whenever an object slides along a straight line relative to another object, their common instant center lies at infinity in either direction perpendicular to the line of motion.

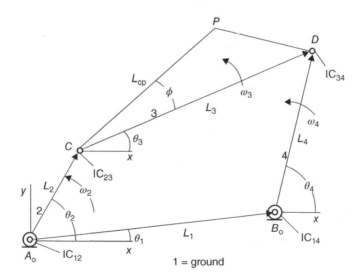

Figure 5.9 Four-bar linkage

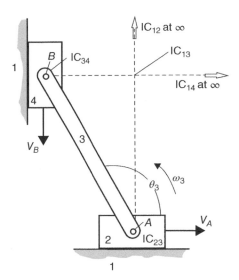

Figure 5.10   Slider–slider linkage

Note that the velocity of a point that is rotating about a fixed point is perpendicular to the line between the rotation center and the point. Therefore, if the direction of the velocity vector is known, the instant center and its center of rotation must lie on a line that is perpendicular to the velocity vector. If the velocity directions of two points on a given object are known, the intersection of the two lines perpendicular to these two velocities will be the center of rotation or the instant center. As the object moves, its center of rotation may move; thus, it is called an instant center. With this in mind, the intersection of the two dashed lines in Figure 5.10 is instant center IC_{13}.

Kennedy's theorem states that any three objects having plane motion relative to one another have their three instant centers along the same straight line. The proof of this theorem is not given here. Seek other references if a proof is desired. Looking back at Figure 5.10, note that IC_{12}, IC_{23}, and IC_{13} are along the same straight line. Also, note that IC_{14}, IC_{34}, and IC_{13} are along the same straight line. Thus, Kennedy's theorem could be used to locate IC_{13} by locating the intersection of these two lines (IC_{12}–IC_{23}) and (IC_{14}–IC_{34}). The intersection point is IC_{13}.

If an object is sliding along a curved path in another object, then the instant center for the two objects is at the center of curvature of the path at that moment. Object 3 is sliding along a curved path in object 2; thus, IC_{23} is located at the center of curvature of the path regardless of whether or not link 2 is moving (see Figure 5.11).

Instant centers consist of two different number subscripts. Instant center 12 and instant center 21 are the same point; thus, the number of possible combinations of two numbers is divided by 2. The number of instant centers (N_{IC}) found in a mechanism is determined using the following equation where n equals the number of links (including the ground link):

$$N_{IC} = \frac{n(n-1)}{2}$$

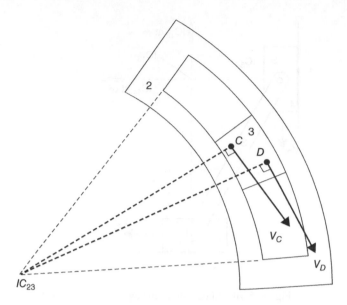

Figure 5.11   Sliding motion

Example 5.4   Locate the instant centers for a 4-bar linkage

Problem: Given the 4-bar linkage drawn to scale in Figure 5.12, locate all instant centers.

Solution: The pin connections are the obvious instant centers; thus, IC_{12}, IC_{23}, IC_{34}, and IC_{14} can be quickly labeled on the 4-bar linkage. Using Kennedy's theorem, IC_{13} is located at the intersection of the line (IC_{12}–IC_{23}) and the line (IC_{14}–IC_{34}). Using Kennedy's theorem, IC_{24} is located at the intersection of the line (IC_{12}–IC_{14}) and the line (IC_{23}–IC_{34}).

Answer: ⬚ See Figure 5.12 ⬚.

Now, look at a slider–crank mechanism.

Example 5.5   Locate the instant centers for a slider–crank mechanism

Problem: Given the slider–crank mechanism drawn to scale in Figure 5.13, locate all instant centers.

Solution: The pin connections are the obvious instant centers; thus, IC_{12}, IC_{23}, and IC_{34} can be quickly labeled on the slider–crank. Since the slider moves along a straight horizontal line, IC_{14} will be on a vertical line located at infinity. Using Kennedy's theorem, IC_{13} is located at the intersection of the line (IC_{12}–IC_{23}) and the line (IC_{14}–IC_{34}). Using Kennedy's theorem, IC_{24} is located at the intersection of the line (IC_{12}–IC_{14}) and the line (IC_{23}–IC_{34}).

Answer: ⬚ See Figure 5.13 ⬚.

When more than four links are involved, it is easier to determine the correct combination of instant centers to use when a helper circle is used to show the already determined instant

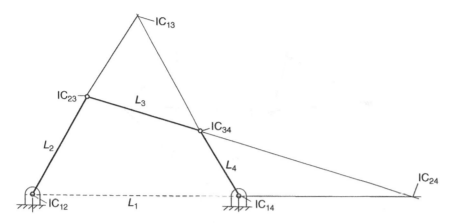

Figure 5.12   Four-bar instant centers

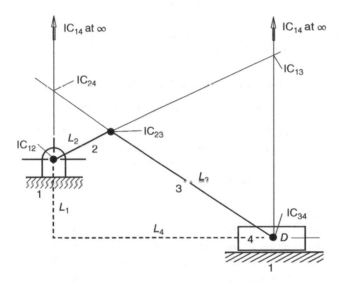

Figure 5.13   Slider–crank instant centers

centers. Since Kennedy's theorem relates three different instant centers and a triangle is made up of three different points, determining which instant center can be solved for next lies in the fact that you can draw two different triangles inside the helper circle with the unknown instant center being a common side to both triangles.

When looking at a 6-bar linkage, there are fifteen different instant centers. $N_{IC} = 6(6-1)/2 = 15$. You create a helper circle by drawing a circle of any size and then evenly spacing six numbers from 1 to 6 around its circumference. For each instant center known, draw a line between the two numbers on the helper circle. For example, when IC_{12} is known, draw a line between the numbers 1 and 2 on the helper circle. Do this for each instant center known.

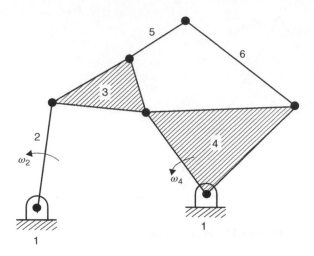

Figure 5.14   Six-bar linkage

If you can draw a line that completes two different triangles in the helper circle, then that instant center can be solved for using Kennedy's theorem.

Example 5.6   Locate the instant centers for a 6-bar linkage

Problem: Given the 6-bar linkage drawn to scale in Figure 5.14, locate all instant centers.

Solution: The pin connections are the obvious instant centers; thus, IC_{12}, IC_{23}, IC_{34}, IC_{14}, IC_{35}, IC_{56}, and IC_{46} can be quickly labeled on the 6-bar linkage. Once this is done, draw the known instant center lines on the helper circle (see Figure 5.15).

Using Kennedy's theorem and the helper circle (see Figure 5.16):

1. IC_{13} is located at the intersection of the line (IC_{12}–IC_{23}) and the line (IC_{14}–IC_{34}).
2. IC_{24} is located at the intersection of the line (IC_{12}–IC_{14}) and the line (IC_{23}–IC_{34}).
3. IC_{45} is located at the intersection of the line (IC_{34}–IC_{35}) and the line (IC_{46}–IC_{56}).
4. IC_{15} is located at the intersection of the line (IC_{14}–IC_{45}) and the line (IC_{13}–IC_{35}).
5. IC_{16} is located at the intersection of the line (IC_{14}–IC_{46}) and the line (IC_{15}–IC_{56}).
6. IC_{26} is located at the intersection of the line (IC_{12}–IC_{16}) and the line (IC_{24}–IC_{46}).
7. IC_{36} is located at the intersection of the line (IC_{34}–IC_{46}) and the line (IC_{35}–IC_{56}).
8. IC_{25} is located at the intersection of the line (IC_{12}–IC_{15}) and the line (IC_{23}–IC_{35}).
9. In addition, as a check of your solution, IC_{25} must be on the line (IC_{26}–IC_{56}).

Answer: $\boxed{\text{See Figure 5.17}}$. There are three lines that go through IC_{25}. The third line is a check.

As an increasing number of instant centers are found, the number of different combinations that can be used to find a new instant center increases. As a result, there are more than just two intersecting lines that determine the location of an instant center. This allows you to check your work as you go along. If the third line does not go through the already determined instant center, then one of the previous instant centers is not located in the proper place on the page.

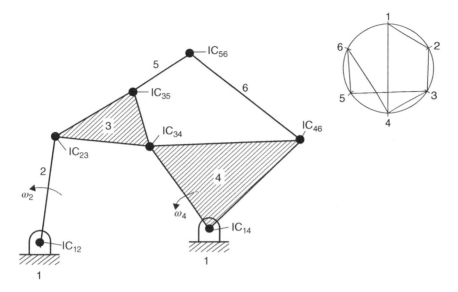

Figure 5.15   Obvious instant centers and helper circle

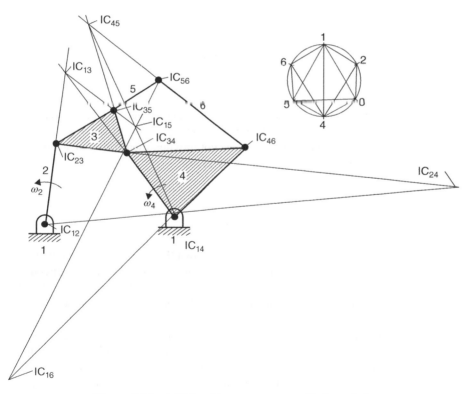

Figure 5.16   Additional instant centers and helper circle

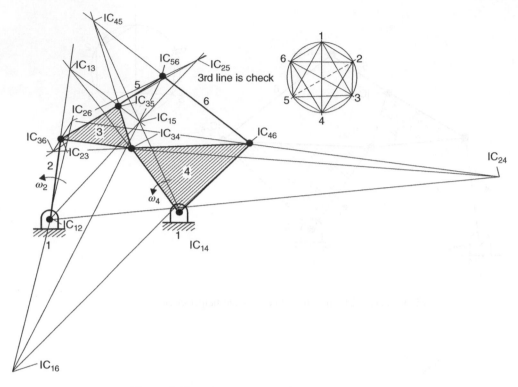

Figure 5.17 All instant center for this 6-bar linkage

When two points are on the same object (Figure 5.18), the relative velocity vector between the two points can be found by multiplying the angular velocity vector of the object by the distance between the two points. If one of the points is not moving, then the relative velocity vector of the other point becomes the absolute velocity vector:

$$\overline{V_{C/A}} = \overline{\omega_{AC} \cdot d_{AC}}$$

$$\overline{V_C} = \overline{V_A} + \overline{\omega_{AC} \cdot d_{AC}}$$

Looking back at a 4-bar linkage with its instant centers shown (Figure 5.19), we note that the velocity vector of IC_{23} can be calculated using the product of ω_2 and d_{23-12} and its direction must be perpendicular to d_{23-12} in the direction of ω_2. Since IC_{12} has a "1" in its subscript, this instant center is not moving; thus, the relative velocity vector V_{23} becomes the absolute velocity vector.

The velocity vector of IC_{23} can be calculated using the product of ω_3 and d_{23-13} and its direction must be perpendicular to d_{23-13} in the direction of ω_3. Since IC_{13} has a "1" in its subscript, this instant center is not moving at this instant; thus, the relative velocity vector V_{23} becomes the absolute velocity vector. Equating the two terms allows us to solve for the angular velocity of link 3:

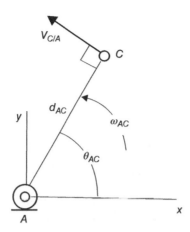

Figure 5.18   Relative velocity

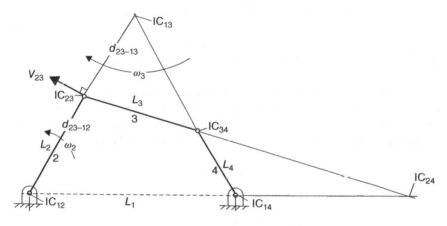

Figure 5.19   Four-bar linkage and velocity of IC$_{23}$

$$\omega_2 \cdot d_{23-12} = V_{23} = \omega_3 \cdot d_{23-13}$$

or

$$\omega_3 = \omega_2 \cdot \left(\frac{d_{23-12}}{d_{23-13}}\right)$$

Note that the angular velocity vector ω_3 must be rotating clockwise if the angular velocity vector ω_2 is rotating counterclockwise (see Figure 5.19). In summary, if the instant centers containing the subscript "1" are on opposite sides of the instant center in common with both points, then the angular velocity vectors must be rotating in opposite directions.

Since the direction of the linear velocity vector for IC$_{23}$ must be perpendicular to d_{23-12} and perpendicular to d_{23-13}, this can only happen if IC$_{12}$, IC$_{23}$, and IC$_{13}$ are along the same straight line. This is what Kennedy's theorem states.

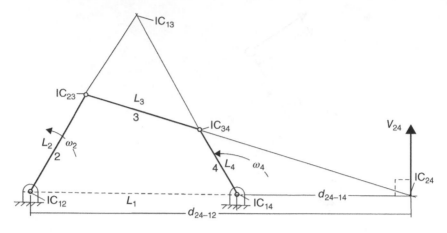

Figure 5.20 Four-bar linkage and velocity of IC$_{24}$

Looking again at a 4-bar linkage with its instant centers already determined (Figure 5.20), the velocity vector of IC$_{24}$ can be calculated using the product of ω_2 and d_{24-12}, and its direction must be perpendicular to d_{24-12} in the direction of ω_2. Since IC$_{12}$ has a "1" in its subscript, this instant center is not moving; thus, the relative velocity vector V_{24} becomes the absolute velocity vector.

The velocity vector of IC$_{24}$ can be calculated using the product of ω_4 and d_{24-14}, and its direction must be perpendicular to d_{24-14} in the direction of ω_4. Since IC$_{14}$ has a "1" in its subscript, this instant center is not moving; thus, the relative velocity vector V_{24} becomes the absolute velocity vector. Equating the two terms allows us to solve for the angular velocity of link 4:

$$\omega_2 \cdot d_{24-12} = V_{24} = \omega_4 \cdot d_{24-14}$$

or

$$\omega_4 = \omega_2 \cdot \left(\frac{d_{24-12}}{d_{24-14}}\right)$$

Note that the angular velocity vector ω_4 must be rotating counterclockwise if the angular velocity vector ω_2 is rotating counterclockwise (see Figure 5.20). In summary, if the instant centers containing the subscript "1" are on the same side of the instant center in common with both points, then the angular velocity vectors must be rotating in the same direction.

Since the direction of the linear velocity vector for IC$_{24}$ must be perpendicular to d_{24-12} and perpendicular to d_{24-14}, this can only happen if IC$_{12}$, IC$_{14}$, and IC$_{24}$ are along the same straight line. This is what Kennedy's theorem states.

The velocity vector of IC$_{34}$ can be calculated using the product of ω_3 and d_{34-13}, and its direction must be perpendicular to d_{34-13} in the direction of ω_3 (see Figure 5.21). Since IC$_{13}$ has a "1" in its subscript, this instant center is not moving at this instant; thus, the relative velocity vector V_{34} becomes the absolute velocity vector. The velocity vector of IC$_{34}$ can be calculated using the product of ω_4 and d_{34-14}, and its direction must be perpendicular to d_{34-14} in the direction of ω_4. Since IC$_{14}$ has a "1" in its subscript, this instant center is not moving; thus, the relative velocity vector V_{34} becomes the absolute velocity vector. Equating the two terms allows us

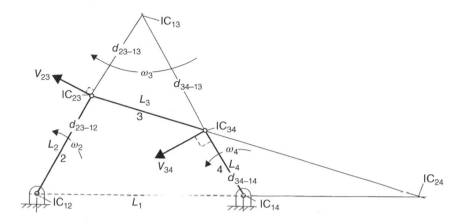

Figure 5.21 Four-bar linkage and velocity of IC_{34}

to solve for the angular velocity of link 4 in terms of ω_3. This is another way to calculate the angular velocity vector for link 4 or a way to check your solution if you have previously solved for ω_4:

$$\omega_3 \cdot d_{34-13} = V_{34} = \omega_4 \cdot d_{34-14}$$

or

$$\omega_4 = \omega_3 \cdot \left(\frac{d_{34-13}}{d_{34-14}}\right)$$

Example 5.7 Use the instant center method to find angular velocity vectors of a 4-bar linkage

Problem: Given the 4-bar linkage drawn to scale in Figure 5.22, with the instant center locations already determined, find the angular velocity vectors for links 3 and 4. Link 2 is 65 mm long and at an angle of 55°. Link 3 is 80 mm long and at 20.8°. Link 4 is 90 mm long and at 114.9°. Link 2 is traveling at 12 rad/s clockwise.

Solution: Relate ω_3 to ω_2 using velocity of IC_{23}. Relate ω_4 to ω_2 using velocity of IC_{24}. Check your solution by relating ω_3 to ω_4 using velocity of IC_{34}. Finally, determine the directions for the two unknown angular velocities:

$$\omega_2 = 12\,\text{rad/s}$$

$$V_{23} = \omega_2 \cdot d_{23-12} = \omega_3 \cdot d_{23-13}$$

$$\omega_3 = \omega_2 \left(\frac{d_{23-12}}{d_{23-13}}\right) = 12 \left(\frac{65.0}{92.2}\right) = 8.460\,\text{rad/s}$$

$$V_{24} = \omega_2 \cdot d_{24-12} = \omega_4 \cdot d_{24-14}$$

$$\omega_4 = \omega_2 \left(\frac{d_{24-12}}{d_{24-14}}\right) = 12 \left(\frac{103.1}{253.1}\right) = 4.888\,\text{rad/s}$$

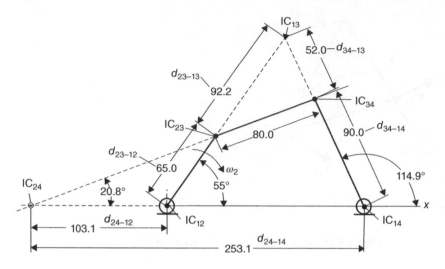

Figure 5.22 Determining angular velocities using instant centers

Checking

$$V_{34} = \omega_3 \cdot d_{34-13} = \omega_4 \cdot d_{34-14}$$

$$\omega_3 = \omega_4 \left(\frac{d_{34-14}}{d_{34-13}} \right) = 4.888 \left(\frac{90.0}{52.0} \right) = 8.460 \, \text{rad/s}$$

Note that ω_2 is rotating clockwise. Since IC_{12} and IC_{13} are on opposite sides of IC_{23}, then ω_3 is rotating the opposite direction from ω_2. Since IC_{12} and IC_{14} are on the same side of IC_{24}, then ω_4 is rotating the same direction as ω_2.

Answer: $\boxed{\overline{\omega_3} = 8.46 \, \text{rad/s c.c.w.}}$ and $\boxed{\overline{\omega_4} = 4.89 \, \text{rad/s c.w.}}$

Previously, we have used the instant center method to calculate the unknown angular velocity vectors. In the case of the slider–crank mechanism shown in Figure 5.23, one of the unknowns is the linear velocity vector of the slider, link 4. Since the slider, link 4, travels along a straight line, every point on link 4 must travel along a straight line. Thus, the linear velocity vector of IC_{24} is also the linear velocity vector of the slider, link 4:

$$\overline{V_{\text{slider}}} = \overline{V_4} = \overline{V_{24}} = \omega_2 \cdot d_{24-12} (\perp d_{24-12})$$

Example 5.8 Use the instant center method to find velocity vectors of a slider–crank mechanism

Problem: Given the slider–crank mechanism drawn to scale in Figure 5.24, with the instant center locations already determined, find the angular velocity vectors for link 3 and the linear velocity vector for link 4. Link 2 is 2.50 in. long and at an angle of 126°. Link 3 is 6.00 in. long. Link 4 is 4.18 in. from the origin. Link 2 is traveling at 14 rad/s clockwise.

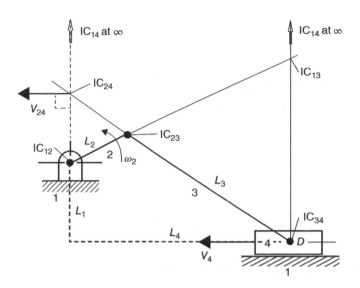

Figure 5.23   Slider–crank mechanism and velocity of link 4

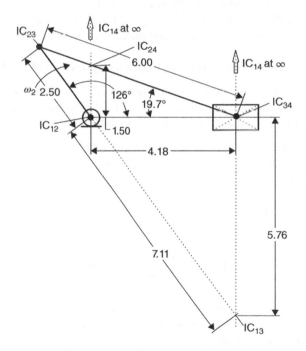

Figure 5.24   Slider–crank velocities

Solution: Relate ω_3 to ω_2 using velocity of IC_{23}. Find the velocity of IC_{24} (also V_4) using ω_2. Check your solution by relating ω_3 to V_4 using the velocity of IC_{34}. Finally, determine the directions for the two unknown velocities:

$$\omega_2 = 14 \, \text{rad/s}$$

$$V_{23} = \omega_2 \cdot d_{23-12} = \omega_3 \cdot d_{23-13}$$

$$\omega_3 = \omega_2 \left(\frac{d_{23-12}}{d_{23-13}} \right) = 14 \left(\frac{2.50}{2.50+7.11} \right) = 3.642 \, \text{rad/s}$$

$$V_4 = V_{24} = \omega_2 \cdot d_{24-12} = 14 \cdot 1.50 = 21.00 \, \text{in./s}$$

Checking

$$\omega_3 \cdot d_{34-13} = V_{34} = V_4$$

$$\omega_3 = \frac{V_{34}}{d_{34-13}} = \frac{21.00}{5.76} = 3.646 \, \text{rad/s (difference due to 3 S.F.s in IC distances)}$$

Note that ω_2 is rotating clockwise. Since IC_{12} and IC_{13} are on the same side of IC_{23}, then ω_3 is rotating the same direction as ω_2. Since V_{24} is moving to the right, V_{slider} is moving to the right.

Answer: $\boxed{\overline{\omega_3} = 3.64 \, \text{rad/s c.w.}}$ and $\boxed{\overline{V_4} = 21.0 \, \text{in./s} \rightarrow}$.

Let us look at the real power of the instant center method. Suppose we were given a 6-bar linkage and we wanted to know the angular velocity vector of link 6 given the angular velocity vector of link 2. We would also like to know the angular direction of link 5. After finding all the instant centers, these two questions can be answered very quickly.

Example 5.9 Use the instant center method to find velocity vectors of a 6-bar linkage

Problem: Given the 6-bar linkage drawn to scale in Figure 5.25, with the instant center locations already determined, find the angular velocity vectors for link 6 and the direction of the angular velocity vector for link 5. Link 2 is traveling at 16 rad/s counter clockwise.

Solution: Relate ω_6 to ω_2 using velocity of IC_{26} and determine the directions for the two unknown angular velocities. Distances are not shown in Figure 5.25 because this would add clutter to an already busy picture:

$$\omega_2 = 16 \, \text{rad/s}$$

$$V_{26} = \omega_2 \cdot d_{26-12} = \omega_6 \cdot d_{26-16}$$

$$\omega_6 = \omega_2 \left(\frac{d_{26-12}}{d_{26-16}} \right) = 16 \left(\frac{104 \, \text{mm}}{261 \, \text{mm}} \right) = 6.375 \, \text{rad/s}$$

Note that ω_2 is rotating counterclockwise. Since IC_{12} and IC_{16} are on the same side of IC_{26}, then ω_6 is rotating the same direction as ω_2. Since IC_{12} and IC_{15} are on the same side of IC_{25}, then ω_5 is rotating the same direction as ω_2.

Answer: $\boxed{\overline{\omega_6} = 6.38 \, \text{rad/s c.c.w.}}$ and $\boxed{\omega_5 = \text{c.c.w.}}$

If you were interested in obtaining all the angular velocity information about this 6-bar linkage, the following equations could be used. Instant center distances are not shown in Figure 5.25:

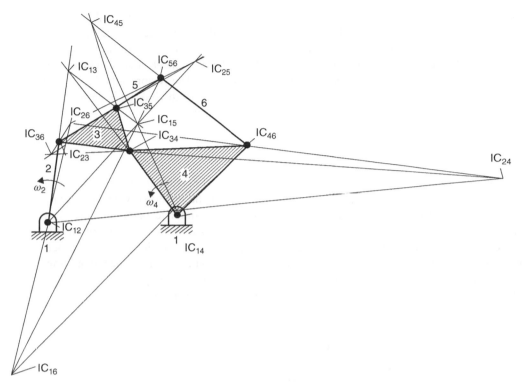

Figure 5.25   Slider–crank velocities

$\omega_2 = 16 \, \text{rad/s (c.c.w.)}$

$V_{23} = \omega_2 \cdot d_{23-12} = \omega_3 \cdot d_{23-13}$

$\omega_3 = \omega_2 \left(\dfrac{d_{23-12}}{d_{23-13}} \right) = 16 \left(\dfrac{80 \, \text{mm}}{72 \, \text{mm}} \right) = 17.78 \, \text{rad/s (c.w.)}$

$V_{24} = \omega_2 \cdot d_{24-12} = \omega_4 \cdot d_{24-14}$

$\omega_4 = \omega_2 \left(\dfrac{d_{24-12}}{d_{24-14}} \right) = 16 \left(\dfrac{450 \, \text{mm}}{323 \, \text{mm}} \right) = 22.29 \, \text{rad/s (c.c.w.)}$

$V_{25} = \omega_2 \cdot d_{25-12} = \omega_5 \cdot d_{25-15}$

$\omega_5 = \omega_2 \left(\dfrac{d_{25-12}}{d_{25-15}} \right) = 16 \left(\dfrac{220 \, \text{mm}}{88 \, \text{mm}} \right) = 40.00 \, \text{rad/s (c.c.w.)}$

$V_{26} = \omega_2 \cdot d_{26-12} = \omega_6 \cdot d_{26-16}$

$\omega_6 = \omega_2 \left(\dfrac{d_{26-12}}{d_{26-16}} \right) = 16 \left(\dfrac{104 \, \text{mm}}{261 \, \text{mm}} \right) = 6.375 \, \text{rad/s (c.c.w.)}$

Checking

$V_{46} = \omega_4 \cdot d_{46-14} = \omega_6 \cdot d_{46-16}$

$\omega_6 = \omega_4 \left(\dfrac{d_{46-14}}{d_{46-16}} \right) = 22.29 \left(\dfrac{93 \, \text{mm}}{325 \, \text{mm}} \right) = 6.378 \, \text{rad/s (c.c.w.)}$

If you wanted to perform a complete velocity analysis of your linkage, the instant center method would not be a good choice because you would need to draw the linkage in each of its configurations, find all the necessary instant center locations, and then perform the simple calculations as shown above. The instant center method is best used to check your analytical solution from another method at a few key orientations of the input link. The vector method described next is the best method to use when a complete velocity analysis of the linkage is desired.

5.4   Vector Method

The vector method uses the equations defined in the analytical positional analysis section of Chapter 2 and calculus to obtain equations that can solve for the unknown velocities at any position of the input link. This method is best used when a complete velocity analysis of the mechanism is desired. The instant center method can be used to check a solution or two of this velocity vector method. Also, if you see a radial jump in the velocity values when performing a complete analysis of the linkage, check the solution on each side of the radial jump. Velocity values tend to follow a pattern and do not jump around radically.

Before continuing, a simple review of some calculus derivatives is in order:

$$\frac{d}{dx}\cos x = -\sin x$$

$$\frac{d}{dx}\sin x = \cos x$$

$$\frac{d}{dx}e^{ax} = ae^{ax} \text{ where "a" is a constant}$$

$$\frac{d}{dt}f(x) = \frac{df(x)}{dx} \cdot \frac{dx}{dt} \text{(chain rule)}$$

The first step when analyzing a 4-bar linkage for velocities is to draw the linkage roughly to scale with vector arrowheads and then determine the unknown position terms. This was done in Chapter 2. Assume all the angular velocity vectors are rotating counterclockwise as seen in Figure 5.26.

Normally, link 2 is the input so ω_2 is known. The angular velocity vectors for links 3 and 4 are the unknowns. Since counterclockwise is the positive direction for angles, you are less likely to make a sign error if you use counterclockwise as positive for the angular velocities as well.

Using vector loop approach from the position analysis of a 4-bar linkage (Figure 2.26) leads to

$$\overline{L_2} + \overline{L_3} = \overline{L_1} + \overline{L_4}$$

Write the equations for the y-component of each vector and then the x-component of each vector:

$$y: \ L_2 \sin\theta_2 + L_3 \sin\theta_3 = L_1 \sin\theta_1 + L_4 \sin\theta_4$$

$$x: \ L_2 \cos\theta_2 + L_3 \cos\theta_3 = L_1 \cos\theta_1 + L_4 \cos\theta_4$$

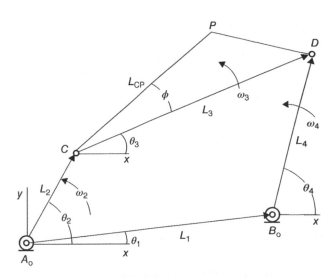

Figure 5.26   Four-bar linkage with vectors

Based on the assumption that the locations of the bearings at A_o and B_o are defined along with the size of links 2, 3, and 4, the two unknowns θ_3 and θ_4 can be determined. Since link 2 is assumed, the input link, its angular position is known, θ_2. Rearranging the two equations so the unknowns are on the left and the two known values are on the right leads to

$$y:\ L_3 \sin\theta_3 - L_4 \sin\theta_4 = L_1 \sin\theta_1 - L_2 \sin\theta_2$$

$$x:\ L_3 \cos\theta_3 - L_4 \cos\theta_4 = L_1 \cos\theta_1 - L_2 \cos\theta_2$$

To obtain the velocity equations, we need to differentiate each equation with respect to time. Note that all the lengths are constants along with θ_1. In addition, ω_2 is known since it assumed to be the driving input link, while ω_1 is zero since the ground is not moving:

$$\frac{dy}{dt}:\ \omega_3 L_3 \cos\theta_3 - \omega_4 L_4 \cos\theta_4 = \omega_1 L_1 \cos\theta_1 - \omega_2 L_2 \cos\theta_2$$

$$\frac{dx}{dt}:\ -\omega_3 L_3 \sin\theta_3 + \omega_4 L_4 \sin\theta_4 = -\omega_1 L_1 \sin\theta_1 + \omega_2 L_2 \sin\theta_2$$

Substitution or Cramer's rule can be used to solve for the unknowns ω_3 and ω_4. The following will use Cramer's rule and determinates to solve the two linear equations containing the two unknowns. The first step is to put the two equations into matrix form as follows:

$$\begin{bmatrix} L_3 \cos\theta_3 & -L_4 \cos\theta_4 \\ -L_3 \sin\theta_3 & L_4 \sin\theta_4 \end{bmatrix} \begin{Bmatrix} \omega_3 \\ \omega_4 \end{Bmatrix} = \begin{Bmatrix} -\omega_2 L_2 \cos\theta_2 \\ \omega_2 L_2 \sin\theta_2 \end{Bmatrix}$$

Apply Cramer's rule using determinates

$$\omega_3 = \cfrac{\begin{bmatrix} -\omega_2 L_2 \cos\theta_2 & -L_4 \cos\theta_4 \\ \omega_2 L_2 \sin\theta_2 & L_4 \sin\theta_4 \end{bmatrix}}{\begin{bmatrix} L_3 \cos\theta_3 & -L_4 \cos\theta_4 \\ -L_3 \sin\theta_3 & L_4 \sin\theta_4 \end{bmatrix}} = \frac{\omega_2 L_2 L_4(-\cos\theta_2 \sin\theta_4 + \sin\theta_2 \cos\theta_4)}{L_3 L_4(\cos\theta_3 \sin\theta_4 - \sin\theta_3 \cos\theta_4)}$$

$$\omega_4 = \cfrac{\begin{bmatrix} L_3 \cos\theta_3 & -\omega_2 L_2 \cos\theta_2 \\ -L_3 \sin\theta_3 & \omega_2 L_2 \sin\theta_2 \end{bmatrix}}{\begin{bmatrix} L_3 \cos\theta_3 & -L_4 \cos\theta_4 \\ -L_3 \sin\theta_3 & L_4 \sin\theta_4 \end{bmatrix}} = \frac{\omega_2 L_3 L_2(\cos\theta_3 \sin\theta_2 - \sin\theta_3 \cos\theta_2)}{L_3 L_4(\cos\theta_3 \sin\theta_4 - \sin\theta_3 \cos\theta_4)}$$

If we use the trigonometry identities listed below, the equations above simplify:

$$\sin(\theta + \phi) = \sin\theta \cos\phi + \cos\theta \sin\phi$$

$$\sin(\theta - \phi) = \sin\theta \cos\phi - \cos\theta \sin\phi$$

$$\boxed{\omega_3 = \frac{\omega_2 L_2 \sin(\theta_2 - \theta_4)}{L_3 \sin(\theta_4 - \theta_3)} = \frac{-\omega_2 L_2 \sin(\theta_2 - \theta_4)}{L_3 \sin(\theta_3 - \theta_4)}} \quad (+\text{c.c.w.}) \qquad (5.1)$$

$$\boxed{\omega_4 = \frac{\omega_2 L_2 \sin(\theta_2 - \theta_3)}{L_4 \sin(\theta_4 - \theta_3)}} \quad (+\text{c.c.w.}) \qquad (5.2)$$

Note the similarity between the two equations above. Link 2's information is in the numerator along with the other unknown's angular position, while the denominator contains the unknown's information and the other unknown's angular position. The negative sign in ω_3's equation indicates a potential change in direction for ω_3 compared to ω_2.

Example 5.10 Use the vector method to find angular velocity vectors of a 4-bar linkage

Problem: Given the 4-bar linkage in Figure 5.27, find the angular velocity vectors for links 3 and 4. Also, determine the linear velocity vector for coupler point, P. Bearing B_o is 350 mm from bearing A_o and at a 10° angle. Link 2 is 150 mm long and at an angle of 60°. Link 3 is 360 mm long and at 27.4°. Link 4 is 240 mm long and at 78.0°. The distance from pin C to coupler point P is 260 mm and point P is located 20° counter clockwise from link 3 (Vector CD). Link 2 is traveling at 18 rad/s counterclockwise.

Solution: Use the two equations found above to solve for the two unknowns, ω_3 and ω_4. Determine the directions for the two unknown angular velocities assuming counterclockwise is positive. Finally, differentiate the position vector for coupler point P to obtain the linear velocity vector for coupler point P:

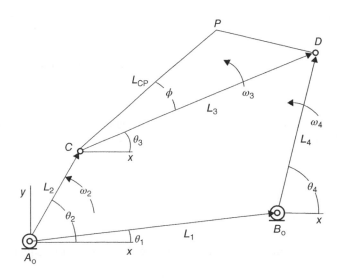

Figure 5.27   Determining angular velocities using vector method

$$\omega_3 = \frac{-\omega_2 L_2 \sin(\theta_2 - \theta_4)}{L_3 \sin(\theta_3 - \theta_4)} = \frac{-(+18)(150 \text{ mm})\sin(60° - 78.0°)}{(360 \text{ mm})\sin(27.4° - 78.0°)} = -2.999$$

$$\overline{\omega_3} = 2.999 \text{ rad/s c.w.}$$

$$\omega_4 = \frac{\omega_2 L_2 \sin(\theta_2 - \theta_3)}{L_4 \sin(\theta_4 - \theta_3)} = \frac{(+18)(150 \text{ mm})\sin(60° - 27.4°)}{(240 \text{ mm})\sin(78.0° - 27.4°)} = 7.844$$

$$\overline{\omega_4} = 7.844 \text{ rad/s c.c.w.}$$

Note that ω_2 is rotating counterclockwise; thus, it is +18 rad/s. Since ω_3 came out negative, it is rotating clockwise. Since ω_4 came out positive, it is rotating counterclockwise.

Now, let's look at the position vector equation for coupler point, P:

$$\overline{P} = (P_x, P_y)$$

$$P_y = L_2 \sin\theta_2 + L_{cp} \sin(\theta_3 + \phi)$$

$$P_x = L_2 \cos\theta_2 + L_{cp} \cos(\theta_3 + \phi)$$

Now, differentiate the vector position equation for coupler point, P:

$$V_{Py} = \frac{dP_y}{dt} = \omega_2 L_2 \cos\theta_2 + \omega_3 L_{cp} \cos(\theta_3 + \phi)$$

$$V_{Py} = (+18)(150 \text{ mm})\cos 60° + (-2.999)(260 \text{ mm})\cos(27.4° + 20°)$$

$$V_{Py} = 822.2 \text{ mm/s}$$

$$V_{Px} = \frac{dP_x}{dt} = -\omega_2 L_2 \sin\theta_2 - \omega_3 L_{cp} \sin(\theta_3 + \phi)$$

$$V_{Px} = -(+18)(150 \text{ mm})\sin 60° - (-2.999)(260 \text{ mm})\sin(27.4° + 20°)$$

$$V_{Px} = -1764 \text{ mm/s}$$

$$\overline{V_P} = (V_{Px}, V_{Py}) = (-1764, 822.2) \text{ mm/s} = 1946 \text{ mm/s} \angle 155°$$

Answer: $\boxed{\omega_3 = 3.00\,\text{rad/s c.w.}}$, $\boxed{\omega_4 = 7.84\,\text{rad/s c.c.w.}}$, and $\boxed{V_P = 1950\,\text{mm/s}\,\angle 155°}$.

Referring again to Figure 5.26, we can write the vector loop equation as follows:

$$\overline{L_2} + \overline{L_3} = \overline{L_1} + \overline{L_4}$$

With the identity shown, one can rewrite the vector loop equation in terms of $e^{i\theta}$. The square root of negative 1 is defined as "i" (some textbooks show this value as "j"):

$$e^{i\theta} = \cos\theta + i \cdot \sin\theta$$

then the vector equation becomes :

$$L_2 e^{i\theta_2} + L_3 e^{i\theta_3} = L_1 e^{i\theta_1} + L_4 e^{i\theta_4}$$

Collecting the imaginary terms and the real terms leads to the same equations determined prior:

$$\text{Img: } i(L_2 \sin\theta_2 + L_3 \sin\theta_3 = L_1 \sin\theta_1 + L_4 \sin\theta_4)$$

$$\text{Real: } L_2 \cos\theta_2 + L_3 \cos\theta_3 = L_1 \cos\theta_1 + L_4 \cos\theta_4$$

If we differentiate the vector equation in its original form, we obtain the velocity equation for the 4-bar linkage directly:

$$\frac{d}{dt}\left(L_2 e^{i\theta_2} + L_3 e^{i\theta_3} = L_1 e^{i\theta_1} + L_4 e^{i\theta_4}\right)$$

$$L_2 i e^{i\theta_2} \dot{\theta}_2 + L_3 i e^{i\theta_3} \dot{\theta}_3 = L_1 i e^{i\theta_1} \dot{\theta}_1 + L_4 i e^{i\theta_4} \dot{\theta}_4$$

Now, substitute the expression for $e^{i\theta} = \cos\theta + i\sin\theta$. Note that $\dot{\theta}$ equals the angular velocity, ω:

$$\dot{\theta}_2 L_2 i(\cos\theta_2 + i \cdot \sin\theta_2) + \dot{\theta}_3 L_3 i(\cos\theta_3 + i \cdot \sin\theta_3) = \dot{\theta}_4 L_4 i(\cos\theta_4 + i \cdot \sin\theta_4)$$

Collect the real and imaginary terms and replace $\dot{\theta}$ with ω. Also, replace i^2 with minus one:

$$\text{Img: } \omega_2 L_2 i \cos\theta_2 + \omega_3 L_3 i \cos\theta_3 = \omega_4 L_4 i \cos\theta_4$$

$$\text{Real: } -\omega_2 L_2 \sin\theta_2 - \omega_3 L_3 \sin\theta_3 = -\omega_4 L_4 \sin\theta_4$$

These are the same two equations we derived earlier when we differentiated the sine and cosine functions. Either by substitution or Cramer's rule, we obtain the following assuming that counterclockwise is positive:

$$\omega_3 = \frac{-\omega_2 L_2 \sin(\theta_2 - \theta_4)}{L_3 \sin(\theta_3 - \theta_4)}$$

$$\omega_4 = \frac{\omega_2 L_2 \sin(\theta_2 - \theta_3)}{L_4 \sin(\theta_4 - \theta_3)}$$

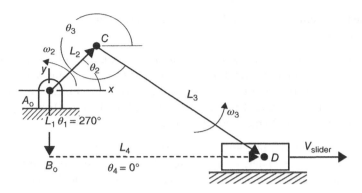

Figure 5.28 Slider–crank velocities

Let us look at a slider–crank mechanism next. Previously, we used the instant center method to calculate the unknown angular velocity vector for link 3 and the linear velocity vector of the slider, link 4. This works well when you are interested in obtaining the velocity information at one particular orientation of the input link. However, if a complete velocity analysis of the slider–crank mechanism is to be done, then the vector loop method is a much better choice (see Figure 5.28). We will use the $e^{i\theta}$ form of the vector equation to solve for the unknown velocities:

$$\overline{L_2} + \overline{L_3} = \overline{L_1} + \overline{L_4}$$

$$L_2 e^{i\theta_2} + L_3 e^{i\theta_3} = L_1 e^{i\theta_1} + L_4 e^{i\theta_4}$$

Differentiate with respect to time and note that θ_3 and L_4 change with respect to time. Link 2 is considered the input link so its angular velocity vector would be known. Link 1 is the ground and does not move. Link 4 only translates so it has no angular velocity:

$$\frac{d}{dt}\left(L_2 e^{i\theta_2} + L_3 e^{i\theta_3} = L_1 e^{i\theta_1} + L_4 e^{i\theta_4}\right)$$

$$iL_2 e^{i\theta_2}\omega_2 + iL_3 e^{i\theta_3}\omega_3 = iL_1 e^{i\theta_1}\cancel{\omega_1} + iL_4 e^{i\theta_4}\cancel{\omega_4} + V_4 e^{i\theta_4}$$

Collect the real and imaginary terms, and then solve for the two unknowns. Note that the equation containing all the imaginary terms has only one unknown present so we can solve for this unknown directly and then use the other equation to solve for the second unknown:

$$i\omega_2 L_2 \cos\theta_2 + i\omega_3 L_3 \cos\theta_3 = iV_4\cancel{\sin\theta_4} = 0$$

$$\boxed{\overline{\omega_3} = \frac{-\omega_2 L_2 \cos\theta_2}{L_3 \cos\theta_3}} \quad (\text{+ indicates c.c.w.}) \tag{5.3}$$

$$i^2\omega_2 L_2 \sin\theta_2 + i^2\omega_3 L_3 \sin\theta_3 = V_4 \cos\theta_4$$

$$\cos\theta_4 = 1 \text{ and } i^2 = -1$$

$$\boxed{V_4 \angle 0° = -\omega_2 L_2 \sin\theta_2 - \omega_3 L_3 \sin\theta_3 = \omega_2 L_2 (\cos\theta_2 \tan\theta_3 - \sin\theta_2)} \tag{5.4}$$

Example 5.11   Use the vector method to find velocity vectors for a slider–crank mechanism

Problem: Given the slider–crank mechanism shown in Figure 5.29, find the angular velocity vector for link 3 and the linear velocity vector for link 4. Link 2 is 2.25 in. long and at an angle of 28°. Link 3 is 6.62 in. long and currently 35.1° below the x-axis. Link 4 is 2.75 in. below the origin traveling along a horizontal line. Currently, the slider, link 4, is 7.403 in. right of the origin at A_o. Link 2 is traveling at 20 rad/s counter clockwise.

Solution: Use the equations derived previously to determine the angular velocity vector, ω_3, and the linear velocity vector of the slider, V_4. Counterclockwise is considered positive in these equations; thus, the angular velocity vector for link 3 is drawn counterclockwise:

$$\omega_3 = \frac{-\omega_2 L_2 \cos\theta_2}{L_3 \cos\theta_3} = \frac{-(+20)(2.25 \text{ in.})\cos 28°}{(6.62 \text{ in.})\cos(-35.1°)} = -7.336$$

$$\overline{\omega_3} = 7.336 \text{ rad/s c.w.}$$

$$V_4 \angle 0° = -\omega_2 L_2 \sin\theta_2 - \omega_3 L_3 \sin\theta_3 = -(+20)(2.25 \text{ in.})\sin 28° - (-7.336)(6.62 \text{ in.})\sin(-35.1°)$$

$$V_4 \angle 0° = -49.05 \text{ in./s}$$

$$\overline{V_4} = 49.05 \text{ in./s} \angle 180°$$

Answer: $\boxed{\overline{\omega_3} = 7.34 \text{ rad/s c.w.}}$ and $\boxed{\overline{V_4} = 49.0 \text{ in./s} \angle 180° \text{ or } \leftarrow}$.

Next, we will consider an inverted slider–crank mechanism shown in Figure 5.30. In this case, the slider moves along link 4 that can rotate about bearing B_o. Both bearings are fixed so that L_1 and θ_1 are constants.

Using B_o as the origin, the vector loop equation becomes

$$\overline{L_3} = \overline{L_1} + \overline{L_2}$$

$$L_3 e^{i\theta_3} = L_1 e^{i\theta_1} + L_2 e^{i\theta_2}$$

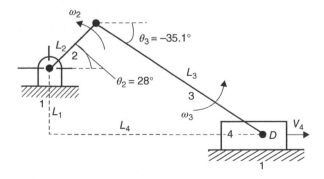

Figure 5.29   Slider–crank mechanism velocities

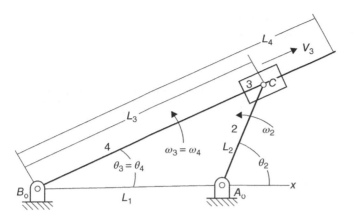

Figure 5.30   Inverted slider–crank

Differentiate with respect to time and note that L_3 changes with time:

$$\frac{d}{dt}\left(L_3 e^{i\theta_3} = L_1 e^{i\theta_1} + L_2 e^{i\theta_2}\right)$$

$$V_{3/4} e^{i\theta_3} + iL_3 e^{i\theta_3}\dot{\theta}_3 = 0 + iL_2 e^{i\theta_2}\dot{\theta}_2$$

Expand the $e^{i\theta}$ terms, and then collect the real and imaginary parts to create two equations:

$$\text{Img: } V_{3/4} i \sin\theta_3 + iL_3\omega_3\cos\theta_3 = iL_2\omega_2\cos\theta_2$$

$$\text{Real: } V_{3/4}\cos\theta_3 + i^2 L_3\omega_3\sin\theta_3 = i^2 L_2\omega_2\sin\theta_2$$

$$i^2 = -1$$

Replace i^2 with minus one and rewrite in matrix form

$$\begin{bmatrix} \sin\theta_3 & L_3\cos\theta_3 \\ \cos\theta_3 & -L_3\sin\theta_3 \end{bmatrix} \begin{Bmatrix} V_{3/4} \\ \omega_3 \end{Bmatrix} = \begin{Bmatrix} L_2\omega_2\cos\theta_2 \\ -L_2\omega_2\sin\theta_2 \end{Bmatrix}$$

Use Cramer's rule to solve for the linear velocity vector of slider 3 along link 4. See equation 5.5.

$$V_{3/4} = \frac{\begin{bmatrix} L_2\omega_2\cos\theta_2 & L_3\cos\theta_3 \\ -L_2\omega_2\sin\theta_2 & -L_3\sin\theta_3 \end{bmatrix}}{\begin{bmatrix} \sin\theta_3 & L_3\cos\theta_3 \\ \cos\theta_3 & -L_3\sin\theta_3 \end{bmatrix}} = \frac{\omega_2 L_2 \cancel{L_3}(-\cos\theta_2\sin\theta_3 + \sin\theta_2\cos\theta_3)}{-\cancel{L_3}(\sin^2\theta_3 + \cos^2\theta_3)}$$

$$V_{3/4} = \frac{\omega_2 L_2(-\cos\theta_2\sin\theta_3 + \sin\theta_2\cos\theta_3)}{-1}$$

$$\boxed{V_{3/4} = \omega_2 L_2\sin(\theta_3 - \theta_2)\angle\theta_3}$$ (5.5)

Use Cramer's rule to solve for the angular velocity of link 3 and link 4 since they have the same angular velocity vector. See equation 5.6.

$$\omega_3 = \frac{\begin{bmatrix} \sin\theta_3 & L_2\omega_2\cos\theta_2 \\ \cos\theta_3 & -L_2\omega_2\sin\theta_2 \end{bmatrix}}{\begin{bmatrix} \sin\theta_3 & L_3\cos\theta_3 \\ \cos\theta_3 & -L_3\sin\theta_3 \end{bmatrix}} = \frac{-L_2\omega_2(\sin\theta_3\sin\theta_2 + \cos\theta_3\cos\theta_2)}{-L_3}$$

$$\boxed{\overline{\omega_3} = \frac{\omega_2 L_2 \cos(\theta_3 - \theta_2)}{L_3}} \quad (+\text{c.c.w.}) \tag{5.6}$$

Example 5.12 **Use the vector method to find velocity vectors for an inverted slider–crank mechanism**

Problem: Given the inverted slider–crank mechanism shown in Figure 5.30, find the angular velocity vector for link 3 and the linear velocity vector for link 3 relative to link 4. Link 2 is 3.25 in. long and at an angle of 65°. Currently, the slider, link 3, is 7.57 in. from bearing B_o and at an angle of 22.9°. Link 2 is traveling at 6 rad/s clockwise.

Solution: Use the equations derived previously. Note that the angular velocities are defined as positive in the counterclockwise direction. The linear velocity vector of the slider on link 4 is positive in the θ_3 direction:

$$\overline{\omega_3} = \frac{\omega_2 L_2 \cos(\theta_3 - \theta_2)}{L_3} \quad (+\text{c.c.w.})$$

$$\overline{\omega_3} = \frac{(-6)(3.25 \text{ in.})\cos(22.9° - 65°)}{7.57 \text{ in.}} = -1.911$$

$$\overline{\omega_3} = 1.911 \text{ rad/s c.w.}$$

$$\overline{V_{3/4}} = \omega_2 L_2 \sin(\theta_3 - \theta_2)\angle\theta_3$$

$$\overline{V_{3/4}} = (-6)(3.25)\sin(22.9° - 65°)\angle 22.9° = +13.07\angle 22.9°$$

$$\overline{V_{3/4}} = 13.07 \text{ in./s} \angle 22.9°$$

Answers: $\boxed{\overline{\omega_3} = 1.91 \text{ rad/s c.w.}}$ and $\boxed{\overline{V_{3/4}} = 13.1 \text{ in./s} \angle 22.9°}$.

Problems

5.1   Given (Figure 5.31): $L_1 = 7.25$ in., $L_2 = 2.00$ in., $L_3 = 7.00$ in., $L_4 = 4.50$ in., $\theta_1 = 5°$, and $\theta_2 = 150°$. A position analysis has determined that the angular positions of links 3 and 4 are $\theta_3 = 27.3°$ and $\theta_4 = 127.4°$. Assuming that link 2 is traveling at 10 rad/s counterclockwise, determine the angular velocity vectors for links 3 and 4.

5.2   Given (Figure 5.31): $L_1 = 185$ mm, $L_2 = 50.0$ mm, $L_3 = 180$ mm, $L_4 = 115$ mm, $\theta_1 = 3°$, and $\theta_2 = 152°$. A position analysis has determined that the angular positions of links 3 and 4 are $\theta_3 = 26.2°$ and $\theta_4 = 125.9°$. Assuming that link 2 is traveling at 12 rad/s counterclockwise, determine the angular velocity vectors for links 3 and 4. (Answers: $\omega_3 = 1.49$ rad/s counter clockwise, $\omega_4 = 4.29$ rad/s counter clockwise.)

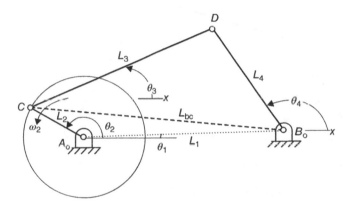

Figure 5.31   Problems 5.1 and 5.2

5.3   Given (Figure 5.32): $L_1 = 8.50$ in., $L_2 = 3.00$ in., $L_3 = 5.00$ in., $L_4 = 4.50$ in., $\theta_1 = 0°$, and $\theta_2 = 60°$. A position analysis has determined that the angular positions of links 3 and 4 are $\theta_3 = 15.5°$ and $\theta_4 = 119.0°$. Assuming that link 2 is traveling at 10 rad/s clockwise, determine the angular velocity vectors for links 3 and 4. (Answers: $\omega_3 = 5.29$ rad/s counter clockwise, $\omega_4 = 4.80$ rad/s clockwise.)

5.4   Given (Figure 5.32): $L_1 = 215$ mm, $L_2 = 75.0$ mm, $L_3 = 125$ mm, $L_4 = 115$ mm, $\theta_1 = 0°$, and $\theta_2 = 60°$. A position analysis has determined that the angular positions of links 3 and 4 are $\theta_3 = 16.1°$ and $\theta_4 = 119.9°$. Assuming that link 2 is traveling at 12 rad/s clockwise, determine the angular velocity vectors for links 3 and 4.

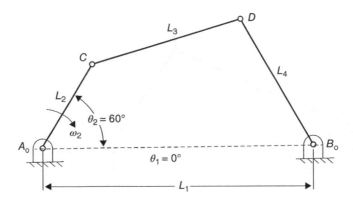

Figure 5.32   Problems 5.3 and 5.4

5.5   Given (Figure 5.33): $L_1 = 7.90$ in., $L_2 = 2.75$ in., $L_3 = 5.40$ in., $L_4 = 4.80$ in., $\theta_1 = 20°$, and $\theta_2 = 310°$. A position analysis has determined that the angular positions of links 3 and 4 are $\theta_3 = 80.6°$ and $\theta_4 = 173.8°$. Assuming that link 2 is traveling at 8 rad/s counterclockwise, determine the angular velocity vectors for links 3 and 4.

5.6   Given (Figure 5.33): $L_1 = 175$ mm, $L_2 = 65.0$ mm, $L_3 = 120$ mm, $L_4 = 105$ mm, $\theta_1 = 20°$, and $\theta_2 = 312°$. A position analysis has determined that the angular positions of links 3 and 4 are $\theta_3 = 82.1°$ and $\theta_4 = 174.1°$. Assuming that link 2 is traveling at 15 rad/s counterclockwise, determine the angular velocity vectors for links 3 and 4. (Answers: $\omega_3 = 5.46$ rad/s counter clockwise, $\omega_4 = 7.11$ rad/s clockwise.)

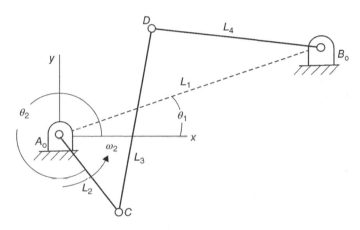

Figure 5.33   Problems 5.5 and 5.6

5.7   Given (Figure 5.34): $L_1 = 7.25$ in., $L_2 = 2.75$ in., $L_3 = 7.25$ in., $L_4 = 3.10$ in., $\theta_1 = 335°$, and $\theta_2 = 98°$, $L_{cp} = 4.75$ in., and $L_{pd} = 2.75$ in. A position analysis has determined that the angular positions of links 3 and 4 are $\theta_3 = 302.3°$ and $\theta_4 = 186.3°$. Assuming that link 2 is traveling at 16 rad/s clockwise, determine the angular velocity vectors for links 3 and 4. Also, determine the linear velocity vector for point P.

5.8   Given (Figure 5.34): $L_1 = 175$ mm, $L_2 = 65.0$ mm, $L_3 = 170$ mm, $L_4 = 75.0$ mm, $\theta_1 = 335°$, and $\theta_2 = 98°$, $L_{cp} = 120$ mm, and $L_{pd} = 70$ mm. A position analysis has determined that the angular positions of links 3 and 4 are $\theta_3 = 303.1°$ and $\theta_4 = 183.2°$. Assuming that link 2 is traveling at 15 rad/s clockwise, determine the angular velocity vectors for links 3 and 4. Also, determine the linear velocity vector for point P. (Answers: $\omega_3 = 6.59$ rad/s counter clockwise, $\omega_4 = 6.36$ rad/s clockwise, $\overline{V}_P = (-487, 495)$ mm/s.)

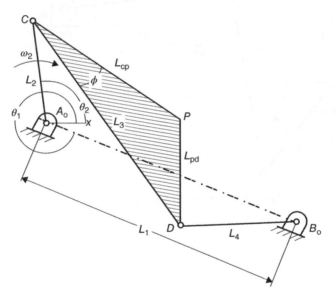

Figure 5.34   Problems 5.7 and 5.8

5.9   Given (Figure 5.35): $L_1 = 3.50$ in., $L_2 = 3.25$ in., $L_3 = 6.50$ in., $\theta_1 = 90°$, $\theta_4 = 0°$, and cur-
        rently $\theta_2 = 110°$. A position analysis has determined that the positions of links 3 and 4
        are $\theta_3 = 3.9°$ and $L_4 = 5.37$ in. Assuming that link 2 is traveling at 4 rad/s counterclock-
        wise, determine the angular velocity vector for link 3 and the linear velocity vector for
        the slider, link 4. (Answers: $\omega_3 = 0.686$ rad/s counter clockwise, $V_4 = 12.5$ in./s←.)
5.10   Given (Figure 5.35): $L_1 = 90.0$ mm, $L_2 = 85.0$ mm, $L_3 = 170$ mm, $\theta_1 = 90°$, $\theta_4 = 0°$, and
        $\theta_2 = 110°$. A position analysis has determined that the positions of links 3 and 4 are $\theta_3 =$
        3.4° and $L_4 = 140.6$ mm. Assuming that link 2 is traveling at 14 rad/s counterclockwise,
        determine the angular velocity vector for link 3 and the linear velocity vector for the
        slider, link 4. (Answers: $\omega_3 = 2.40$ rad/s counter clockwise, $V_4 = 1140$ mm/s←.)

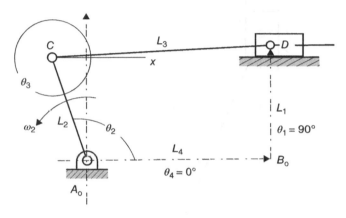

Figure 5.35   Problems 5.9 and 5.10

5.11 Given (Figure 5.36): $L_2 = 2.25$ in., $L_3 = 5.25$ in., $\theta_4 = 90°$, and currently $\theta_2 = 32°$. A position analysis has determined that the positions of links 3 and 4 are $\theta_3 = 111.3°$ and $L_4 = 6.08$ in. Assuming that link 2 is traveling at 4 rad/s clockwise, determine the angular velocity vector for link 3 and the linear velocity vector for the slider, link 4.

5.12 Given (Figure 5.36): $L_2 = 56.0$ mm, $L_3 = 130$ mm, $\theta_4 = 90°$, and $\theta_2 = 32°$. A position analysis has determined that the positions of links 3 and 4 are $\theta_3 = 111.4°$ and $L_4 = 150.7$ mm. Assuming that link 2 is traveling at 14 rad/s clockwise, determine the angular velocity vector for link 3 and the linear velocity vector for the slider, link 4. (Answers: $\omega_3 = 3.43$ rad/s counter clockwise, $V_4 = 828$ mm/s↓.)

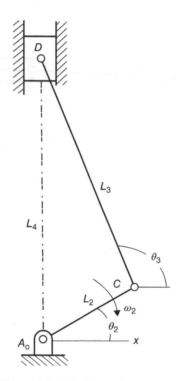

Figure 5.36 Problems 5.11 and 5.12

5.13 Given (Figure 5.37): $L_1 = 1.50$ in., $L_2 = 2.25$ in., $L_3 = 5.25$ in., $\theta_1 = 118°$, $\theta_4 = 28°$, and currently $\theta_2 = 88°$. A position analysis has determined that the positions of links 3 and 4 are $\theta_3 = 23.1°$ and $L_4 = 6.36$ in. Assuming that link 2 is traveling at 7 rad/s clockwise, determine the angular velocity vector for link 3 and the linear velocity vector for the slider, link 4.

5.14 Given (Figure 5.37): $L_1 = 40.0$ mm, $L_2 = 62.0$ mm, $L_3 = 140$ mm, $\theta_1 = 118°$, $\theta_4 = 28°$, and currently $\theta_2 = 90°$. A position analysis has determined that the positions of links 3 and 4 are $\theta_3 = 22.0°$ and $L_4 = 168.3$ mm. Assuming that link 2 is traveling at 18 rad/s clockwise, determine the angular velocity vector for link 3 and the linear velocity vector for the slider, link 4.

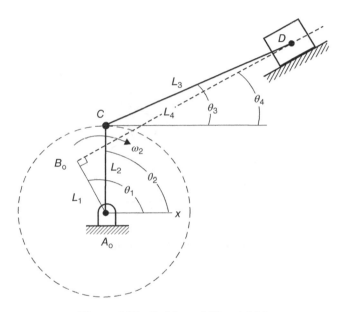

Figure 5.37 Problems 5.13 and 5.14

5.15 Given (Figure 5.38): $L_1 = 5.60$ in., $L_2 = 3.25$ in., $L_4 = 10.0$ in., $\theta_1 = 0$, $\theta_4 = \theta_3$, and currently $\theta_2 = 66°$. A position analysis has determined that the positions of links 3 and 4 are $\theta_4 = 23.2°$ and $L_3 = 7.53$ in. Assuming that link 2 is traveling at 4 rad/s counterclockwise, determine the angular velocity vector for link 4 and the linear velocity vector of the slider, link 3, relative to link 4.

5.16 Given (Figure 5.38): $L_1 = 140$ mm, $L_2 = 85.0$ mm, $L_4 = 250$ mm, $\theta_1 = 0°$, $\theta_4 = \theta_3$, and currently $\theta_2 = 66°$. A position analysis has determined that the positions of links 3 and 4 are $\theta_4 = 24.0°$ and $L_3 = 191.1$ mm. Assuming that link 2 is traveling at 14 rad/s counterclockwise, determine the angular velocity vector for link 4 and the linear velocity vector of the slider, link 3, relative to link 4.

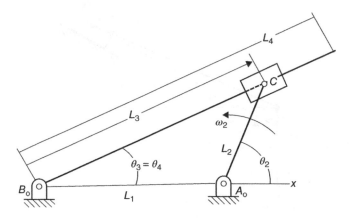

Figure 5.38 Problems 5.15 and 5.16

5.17   Given (Figure 5.39): $L_1 = 2.88$ in., $L_2 = 2.75$ in., $L_3 = 4.00$ in., $L_4 = 3.25$ in., $L_5 = 4.75$ in., $L_6 = 4.00$ in., $L_7 = 4.38$ in., $\theta_1 = \theta_7 = 0°$, and currently $\theta_2 = 101°$ and currently rotating at 7.00 rad/s counter clockwise. Also, $L_{cp} = 2.50$ in., $L_{pd} = 2.37$ in., $L_{P1P2} = 3.25$ in., and $L_{P2F} = 2.38$ in. A positional analysis has determined that the angular positions of links 3, 4, 5, and 6 are $\theta_3 = 7.2°$, $\theta_4 = 80.0°$, $\theta_5 = -6.2°$, and $\theta_6 = 107.1°$. Similarly, the current locations of point P_1 and point P_2 are $P_1 = (1.36, 4.34)$ in. and $P_2 = (4.40, 5.50)$ in. Find the angular velocity vectors for links 3, 4, 5, and 6. Also, find the current linear velocity vectors for points P_1 and P_2.

5.18   Given (Figure 5.39): $L_1 = 75.0$ mm, $L_2 = 70.0$ mm, $L_3 = 100$ mm, $L_4 = 80.0$ mm, $L_5 = 120$ mm, $L_6 = 100$ mm, $L_7 = 115$ mm, $\theta_1 = \theta_7 = 0°$, and currently $\theta_2 = 95°$ and currently rotating at 9.00 rad/s counter clockwise. Also, $L_{cp} = 65.0$ mm, $L_{pd} = 60.0$ mm, $L_{P1P2} = 85.0$ mm, and $L_{P2F} = 60.0$ mm. A positional analysis has determined that the angular positions of links 3, 4, 5, and 6 are $\theta_3 = 4.6°$, $\theta_4 = 76.6°$, $\theta_5 = -7.2°$, and $\theta_6 = 105.7°$. Similarly, the current locations of point P_1 and point P_2 are $P_1 = (43.8, 111)$ mm and $P_2 = (123, 141)$ mm. Find the angular velocity vectors for links 3, 4, 5, and 6. Also, find the current linear velocity vectors for points P_1 and P_2.

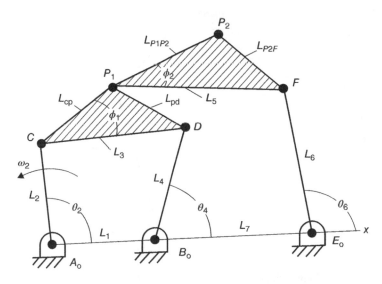

Figure 5.39   Problems 5.17 and 5.18

5.19   Given (Figure 5.40): $L_1 = 3.00$ in., $L_2 = 1.55$ in., $L_3 = 2.50$ in., $L_4 = 2.75$ in., $L_5 = 2.88$ in., $L_6 = 3.88$ in., $L_7 = 3.25$ in., $\theta_1 = 180°$, $\theta_7 = -44°$, and currently $\theta_2 = 103°$ and travel-ing at 8 rad/s counter clockwise. Also, $L_{cp} = 2.88$ in. and $L_{pd} = 4.25$ in. A positional anal-ysis has determined that the angular positions of links 3, 4, 5, and 6 are $\theta_3 = 151.3°$, $\theta_4 = 80.4°$, $\theta_5 = 130.9°$, and $\theta_6 = 72.6°$. Similarly, the current location of point P is (4.61, 3.62) in. Find the angular velocity vectors for links 3, 4, 5, and 6. Also, find the current linear velocity vector for point P.

5.20   Given (Figure 5.40): $L_1 = 75.0$ mm, $L_2 = 40.0$ mm, $L_3 = 60.0$ mm, $L_4 = 70.0$ mm, $L_5 = 72.0$ mm, $L_6 = 95.0$ mm, $L_7 = 80.0$ mm, $\theta_1 = 180°$, $\theta_7 = -43°$, and currently $\theta_2 = 102°$ and currently traveling at 12 rad/s counter clockwise. Also, $L_{cp} = 72.0$ mm and $L_{pd} = 105$ mm. A positional analysis has determined that the angular positions of links 3, 4, 5, and 6 are $\theta_3 = 150.7°$, $\theta_4 = 78.2°$, $\theta_5 = 130.2°$, and $\theta_6 = 71.6°$. Similarly, the current location of point P is (117, 90.6) mm. Find the angular velocity vectors for links 3, 4, 5, and 6. Also, find the current linear velocity vector for point P.

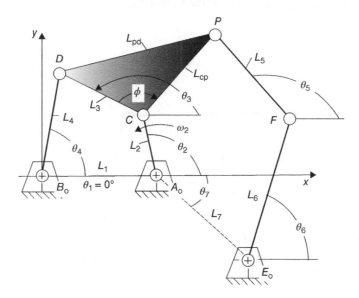

Figure 5.40   Problems 5.19 and 5.20

Programming Exercises

P5.1   Given (Figure 5.41): $L_1 = 7.25$ in., $L_2 = 2.00$ in., $L_3 = 7.00$ in., $L_4 = 4.50$ in., and $\theta_1 = 5°$.
Start with $\theta_2 = 5°$ and increment it by $10°$ for a complete revolution. Link 2 is traveling at
a constant 10 rad/s counterclockwise. Determine the angular velocity vectors for links 3
and 4 for each θ_2 value.

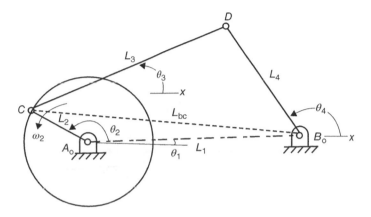

Figure 5.41   Programming exercise P5.1

P5.2   Given: $L_1 = 3.52$ in., $L_2 = 1.25$ in., $L_3 = 4.00$ in., $L_4 = 2.00$ in., $\theta_1 = 0°$, $L_{cp} = 3.00$ in., and
$L_{pd} = 2.00$ in. Assume link 2 is traveling at a constant 12 rad/s counter clockwise. Start with
$\theta_2 = 0°$ and increment it by $5°$ for a complete revolution. Find the angular velocity vectors
for links 3 and 4 assuming Figure 5.42 is drawn roughly to scale. Also, determine the velocity vector of point P. Plot magnitude of the velocity of point P versus the input angle θ_2.

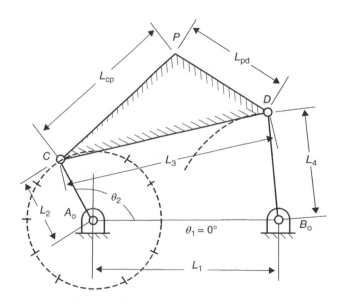

Figure 5.42   Programming exercise P5.2

P5.3   Given: $L_1 = 70.0$ mm, $L_2 = 55.0$ mm, $L_3 = 165$ mm, $\theta_1 = 270°$, and $\theta_4 = 0°$. Assume link
       2 is traveling at a constant 20 rad/s counter clockwise. Start with $\theta_2 = -18.5°$ and incre-
       ment it by 10° for a complete revolution. Find the angular velocity vector for link 3 and
       the linear velocity vector for link 4 (point D) assuming Figure 5.43 is drawn roughly
       to scale.

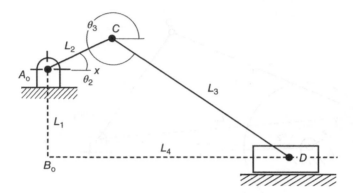

Figure 5.43   Programming exercise P5.3

P5.4   Given: $L_1 = 30.0$ mm, $L_2 = 55.0$ mm, $L_3 = 210$ mm, $\theta_1 = 270°$, $\theta_4 = 0°$, $L_{cp} = 84.0$ mm,
       and $L_{pd} = 170$ mm. Assume link 2 is traveling at a constant 10 rad/s clockwise. Start with
       $\theta_2 = 360°$ and decrement it by 5° for a complete revolution. Find the angular velocity
       vector of link 3 and the linear velocity vector of link 4 (point D) assuming
       Figure 5.44 is drawn roughly to scale. Also, find the linear velocity vector (V_{px}, V_{py})
       for point P for a complete revolution of link 2.

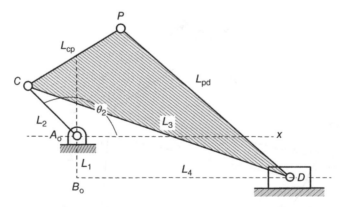

Figure 5.44   Programming exercise P5.4

P5.5   Given: $L_1 = 4.00$ in., $L_2 = 2.50$ in., $L_5 = 3.75$ in., $\theta_1 = \theta_5 = 0°$, and $\theta_3 = \theta_4$. Link 2 is traveling at a constant 16 rad/s counter clockwise. Start with $\theta_2 = 0°$ and increment it by 10°
for a complete revolution. Find the angular velocity vector for link 4, the linear velocity
vector of L_3 along link 4 (B_o to point C), and the vertical velocity vector of point D
assuming Figure 5.45 is drawn roughly to scale. Plot the vertical velocity vector of point
D versus the input angle θ_2.

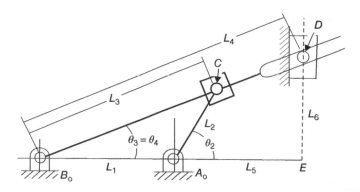

Figure 5.45   Programming exercise P5.5

P5.6   Given: $L_1 = 3.00$ in., $L_2 = 1.38$ in., $L_3 = 2.50$ in., $L_4 = 2.75$ in., $L_5 = 4.75$ in., $L_6 = 4.88$ in.,
$L_7 = 3.25$ in., $\theta_1 = 180°$, $\theta_7 = -44°$, $L_{cp} = 2.88$ in., and $L_{pd} = 4.25$ in. Link 2 is traveling at
a constant 10 rad/s counter clockwise. Start with $\theta_2 = 0°$ and increment it by 12° for a
complete revolution. Find the angular velocity vectors for links 3, 4, 5, and 6 assuming
Figure 5.46 is drawn roughly to scale.

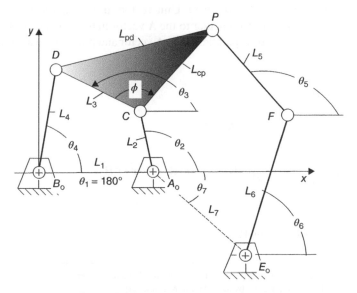

Figure 5.46   Programming exercise P5.6

6

Acceleration

6.1 Introduction

Acceleration analysis is important if a dynamic force analysis is to be performed. Acceleration in linkages is important because inertia forces are proportional to rectilinear acceleration and inertia torques are proportional to angular accelerations. Inertia forces influence stresses in the parts of the machine, the bearing loads, vibration, and noise. And acceleration analysis of a mechanism is done by adding relative accelerations. This method is similar to the velocity method that uses relative velocities.

A vector is defined as $\bar{\mathbf{A}} = \mathbf{R} \angle \theta = \mathbf{R} e^{i\theta} = \mathbf{R} cos\theta + i\mathbf{R} sin\theta$. If the vector is differentiated twice with respect to time, we obtain the acceleration vector. θ, ω, and α are assumed positive when rotating counterclockwise (c.c.w.) (see Figure 6.1). Unit vectors $\mathbf{u}_t$ and $\mathbf{u}_n$ are shown on the figure for reference. Unit vector $\mathbf{u}_t$ is perpendicular to the $\mathbf{A}$ vector in the direction of the angular acceleration vector, α. Unit vector $\mathbf{u}_n$ is parallel to the $\mathbf{A}$ vector and pointed toward the center of curvature:

$$\bar{\mathbf{A}} = \mathbf{R} \cdot e^{i\theta}$$

$$\bar{v}_A = \frac{d}{dt}\left(\mathbf{R} \cdot e^{i\theta}\right) = i\omega \mathbf{R} \cdot e^{i\theta}$$

$$\bar{a}_A = \frac{d}{dt}\left(\bar{v}_A\right) = \frac{d}{dt}\left(i\omega \mathbf{R} \cdot e^{i\theta}\right) = i\alpha \mathbf{R} \cdot e^{i\theta} + i^2\omega^2 \mathbf{R} \cdot e^{i\theta}$$

$$i^2 = -1$$

$$\bar{a}_A = i\alpha \mathbf{R} \cdot e^{i\theta} - \omega^2 \mathbf{R} \cdot e^{i\theta}$$

Design and Analysis of Mechanisms: A Planar Approach, First Edition. Michael J. Rider.
© 2015 John Wiley & Sons, Ltd. Published 2015 by John Wiley & Sons, Ltd.

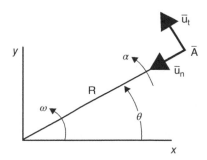

Figure 6.1   Rotating vector

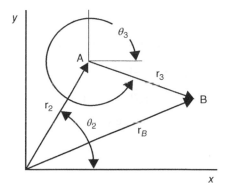

Figure 6.2   Relative vectors

$$\bar{\mathbf{u}}_t = ie^{i\theta}$$

$$\bar{\mathbf{u}}_n = -e^{i\theta}$$

$$\bar{a}_\mathbf{A} = (\alpha\mathbf{R})\bar{\mathbf{u}}_t + (\omega^2\mathbf{R})\bar{\mathbf{u}}_n$$

or

$$\bar{a}_\mathbf{A} = \alpha\mathbf{R}\angle(\theta + 90°) + \omega^2\mathbf{R}\angle(\theta + 180°)$$

6.2   Relative Acceleration

Relative acceleration is the acceleration of one point relative to another point. It can be derived by starting with relative position vectors and then differentiating twice with respect to time. There are two ways to get from the origin to point B: going to point A using r_2 and then to point B using r_3 or going directly to point B using r_B (see Figure 6.2). Thus, $\overline{r_2} + \overline{r_3} = \overline{r_B}$ or $\overline{r_A} + \overline{r_{B/A}} = \overline{r_B}$.

If we differentiate twice with respect to time, we obtain the following:

$$\frac{d}{dt}\left(\overline{r_A}+\overline{r_{B/A}}=\overline{r_B}\right) \text{ leads to: } \overline{v_A}+\overline{v_{B/A}}=\overline{v_B}$$

$$\frac{d}{dt}\left(\overline{v_A}+\overline{v_{B/A}}=\overline{v_B}\right) \text{ leads to: } \overline{a_A}+\overline{a_{B/A}}=\overline{a_B}$$

If the distance from the origin to point A is fixed and the distance between A and B is fixed as it would be for two fixed size links of a mechanism, then the acceleration of B can be determined. Note that point A travels in a circular path around the origin; thus, it has a normal and a tangential acceleration. Similarly, point B travels in a circular path around point A; thus, it has a normal and a tangential acceleration relative to point A. Counterclockwise rotation is assumed to be positive:

$$\overline{a_B}=\overline{a_{An}}+\overline{a_{At}}+\overline{a_{B/An}}+\overline{a_{B/At}}$$

where

$$\overline{a_{An}}=\omega_2{}^2 r_2\angle(\theta_2+180°)$$

$$\overline{a_{At}}=\alpha_2 r_2\angle(\theta_2+90°)$$

$$\overline{a_{B/An}}=\omega_3{}^2 r_3\angle(\theta_3+180°)$$

$$\overline{a_{B/At}}=\alpha_3 r_3\angle(\theta_3+90°)$$

6.3 Slider–Crank Mechanism with Horizontal Motion

With this in mind, we are now ready to perform an acceleration analysis on a slider–crank mechanism. To perform an acceleration analysis on a slider–crank mechanism, we first write the position equation based on a vector loop and then differentiate it twice with respect to time (see Figure 6.3). The input link is link two, so θ_2, ω_2, and α_2 are known:

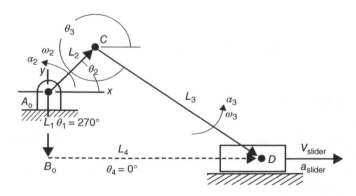

Figure 6.3 Slider–crank mechanism with horizontal motion

$$\overline{L_2} + \overline{L_3} - \overline{L_4} - \overline{L_1} = 0$$

$$L_2 \angle \theta_2 + L_3 \angle \theta_3 - L_4 \angle \theta_4 - L_1 \angle \theta_1 = 0$$

$$L_2 e^{i\theta_2} + L_3 e^{i\theta_3} - L_4 e^{i\theta_4} - L_1 e^{i\theta_1} = 0$$

Differentiating with respect to time generates the velocity loop equation:

$$L_2 i \omega_2 e^{i\theta_2} + L_3 i \omega_3 e^{i\theta_3} - v_4 e^{i\theta_4} - L_4 i \not\omega_4 e^{i\theta_4} - L_1 i \not\omega_1 e^{i\theta_1} = 0$$

Differentiating with respect to time again generates the acceleration loop equation:

$$L_2 i \alpha_2 e^{i\theta_2} - L_2 \omega_2^2 e^{i\theta_2} + L_3 i \alpha_3 e^{i\theta_3} - L_3 \omega_3^2 e^{i\theta_3} - a_4 e^{i\theta_4} = 0$$

Substituting $e^{i\theta} = \cos\theta + i \sin\theta$ and collecting the real and imaginary terms lead to

$$L_2 i \alpha_2 (\cos\theta_2 + i \sin\theta_2) - L_2 \omega_2^2 (\cos\theta_2 + i \sin\theta_2) + L_3 i \alpha_3 (\cos\theta_3 + i \sin\theta_3)$$

$$- L_3 \omega_3^2 (\cos\theta_3 + i \sin\theta_3) - a_4 (\cos\theta_4 + i \sin\theta_4) = 0$$

real: $-L_2 \alpha_2 \sin\theta_2 - L_2 \omega_2^2 \cos\theta_2 - L_3 \alpha_3 \sin\theta_3 - L_3 \omega_3^2 \cos\theta_3 - a_4 \cos\theta_4 = 0$

img: $+ L_2 \alpha_2 \cos\theta_2 - L_2 \omega_2^2 \sin\theta_2 + L_3 \alpha_3 \cos\theta_3 - L_3 \omega_3^2 \sin\theta_3 - a_4 \not\sin\theta_4 = 0$

The equation containing all the imaginary terms has only one unknown, α_3, so it can be used to solve for the angular acceleration of link 3. The equation containing all the real terms can be used to solve for the linear acceleration of the slider, a_4. See equations 6.1, 6.2, and 6.3. Note that $\cos\theta_4 = 1$:

$$\boxed{\alpha_3 = \frac{L_2 \omega_2^2 \sin\theta_2 - L_2 \alpha_2 \cos\theta_2 + L_3 \omega_3^2 \sin\theta_3}{L_3 \cos\theta_3}} \quad (+\text{c.c.w}) \tag{6.1}$$

$$\boxed{a_4 \angle 0° = a_{\text{slider}} = -L_2 \alpha_2 \sin\theta_2 - L_2 \omega_2^2 \cos\theta_2 - L_3 \alpha_3 \sin\theta_3 - L_3 \omega_3^2 \cos\theta_3} \tag{6.2}$$

If α_3 is substituted into the a_4 equation, the linear acceleration of the horizontal slider simplifies to the following. Note that θ_3 cannot become vertical without causing a divide by zero:

$$\boxed{a_4 \angle 0° = L_2 \alpha_2 (\cos\theta_2 \tan\theta_3 - \sin\theta_2) - L_2 \omega_2^2 (\sin\theta_2 \tan\theta_3 + \cos\theta_2) - \frac{L_3 \omega_3^2}{\cos\theta_3}} \tag{6.3}$$

From the position analysis for a slider–crank mechanism with horizontal motion,

$$\boxed{\begin{aligned} \theta_3 &= \sin^{-1}\left(\frac{-L_1 - L_2 \sin\theta_2}{L_3}\right) \\ L_4 &= L_2 \cos\theta_2 + L_3 \cos\theta_3 \end{aligned}} \tag{2.4} \tag{2.5}$$

From the velocity analysis for a slider–crank mechanism with horizontal motion,

$$\omega_3 = \frac{-\omega_2 L_2 \cos\theta_2}{L_3 \cos\theta_3} \quad (+\text{c.c.w.}) \tag{5.3}$$

$$V_4\angle 0° = V_{\text{slider}} = -\omega_2 L_2 \sin\theta_2 - \omega_3 L_3 \sin\theta_3 = \omega_2 L_2(\cos\theta_2 \tan\theta_3 - \sin\theta_2) \tag{5.4}$$

Example 6.1 Acceleration analysis of a slider–crank mechanism

Problem: A slider–crank linkage (Figure 6.4) has link lengths of $L_1 = 5.62$ in., $L_2 = 4.00$ in., and $L_3 = 5.00$ in. Bearing A_o is located at the origin. For the current position of $\theta_2 = 240°$, the angle $\theta_3 = -25.5°$ and the length $L_4 = 2.51$ in. Note $\theta_1 = 270°$ and $\theta_4 = 0°$. Link 2 is currently traveling at 7.00 rad/s counter clockwise and accelerating at a rate of 1.50 rad/s^2. Determine the angular acceleration vector for link 3 (α_3) and the linear acceleration vector for the slider (a_{slider}).

Solution:

1. Verify the position vectors given.
2. Calculate the velocity vectors.
3. Calculate the acceleration vectors.

Position analysis verification:

$$\theta_3 = \sin^{-1}\left(\frac{-L_1 - L_2 \sin\theta_2}{L_3}\right) = \sin^{-1}\left(\frac{-5.62 - 4.00\sin 240°}{5.00}\right) = \sin^{-1}(-0.4312) = -25.54°$$

$$L_4 = L_2 \cos\theta_2 + L_3 \cos\theta_3 = 4.00\cos 240° + 5.00\cos(-25.54°) = 2.511 \text{ in.}\angle 0°$$

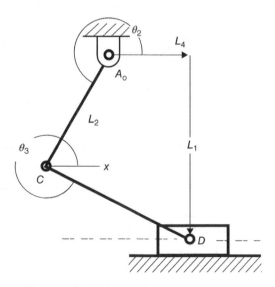

Figure 6.4 Slider–crank acceleration analysis

Velocity analysis:

$$\omega_3 = \frac{-\omega_2 L_2 \cos\theta_2}{L_3 \cos\theta_3} = \frac{-7.00(4.00)\cos 240°}{5.00\cos(-25.54°)} = 3.103 \text{ rad/s c.c.w.}$$

$$V_{\text{slider}}\angle 0° = -\omega_2 L_2 \sin\theta_2 - \omega_3 L_3 \sin\theta_3 = -7.00(4.00)\sin 240° - 3.103(5.00)\sin(-25.54°)$$

$$V_{\text{slider}} = 30.94 \text{ in./s} \angle 0°$$

Acceleration analysis:

$$\alpha_3 = \frac{L_2\omega_2^2 \sin\theta_2 - L_2\alpha_2 \cos\theta_2 + L_3\omega_3^2 \sin\theta_3}{L_3 \cos\theta_3}$$

$$\alpha_3 = \frac{4.00(7.00)^2 \sin 240° - 4.00(-1.50)\cos 240° + 5.00(3.103)^2 \sin(-25.54°)}{5.00\cos(-25.54°)}$$

$$\alpha_3 = \frac{-193.50}{4.511} = -42.89 \text{ rad/s}^2$$

$$a_{\text{slider}}\angle 0° = -L_2\alpha_2 \sin\theta_2 - L_2\omega_2^2 \cos\theta_2 - L_3\alpha_3 \sin\theta_3 - L_3\omega_3^2 \cos\theta_3$$

$$a_{\text{slider}}\angle 0° = -4.00(-1.50)\sin(240°) - 4.00(7.00)^2 \cos(240°) - 5.00(-42.89)\sin(-25.54°)\ldots$$

$$-5.00(3.103)^2 \cos(-25.54°)$$

$$a_{\text{slider}}\angle 0° = -43.1 \text{ in./s}^2$$

Answers: $\boxed{\alpha_3 = 42.9 \text{ rad/s}^2 \text{ c.w.}}$ and $\boxed{a_{\text{slider}} = 43.1 \text{ in./s}^2 \angle 180°}$.

6.4 Acceleration of Mass Centers for Slider–Crank Mechanism

To perform a dynamic force analysis, the accelerations of the mass centers of all moving links must be determined. If we assume the links are made from homogeneous material, then their mass centers will be located at their geometric centers (see Figure 6.5) where G_2, G_3, and G_4 represent the mass centers of links 2, 3, and 4.

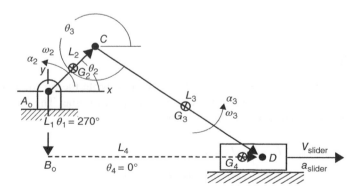

Figure 6.5 Mass centers for slider–crank

Mass center G_2 is located at the midpoint of link 2; thus,

$$\overline{L_{G_2}} = \frac{1}{2}\overline{L_2} = \frac{1}{2}L_2 e^{i\theta_2}$$

Differentiating twice with respect to time leads to

$$\overline{a_{G_2}} = \frac{1}{2}L_2 i\alpha_2 e^{i\theta_2} + \frac{1}{2}L_2 i^2 \omega_2^2 e^{i\theta_2}$$

$$a_{xG_2} = -\frac{1}{2}L_2 \alpha_2 \sin\theta_2 - \frac{1}{2}L_2 \omega_2^2 \cos\theta_2$$

$$a_{yG_2} = \frac{1}{2}L_2 \alpha_2 \cos\theta_2 - \frac{1}{2}L_2 \omega_2^2 \sin\theta_2$$

Mass center G_3 is located at the midpoint of link 3; thus,

$$\overline{L_{G_3}} = \overline{L_2} + \frac{1}{2}\overline{L_3} = L_2 e^{i\theta_2} + \frac{1}{2}L_3 e^{i\theta_3}$$

Differentiating twice with respect to time leads to

$$\overline{a_{G_3}} = L_2 i\alpha_2 e^{i\theta_2} + L_2 i^2 \omega_2^2 e^{i\theta_2} + \frac{1}{2}L_3 i\alpha_3 e^{i\theta_3} + \frac{1}{2}L_3 i^2 \omega_3^2 e^{i\theta_3}$$

$$a_{xG_3} = -L_2 \alpha_2 \sin\theta_2 - L_2 \omega_2^2 \cos\theta_2 - \frac{1}{2}L_3 \alpha_3 \sin\theta_3 - \frac{1}{2}L_3 \omega_3^2 \cos\theta_3$$

$$a_{yG_3} = L_2 \alpha_2 \cos\theta_2 - L_2 \omega_2^2 \sin\theta_2 + \frac{1}{2}L_3 \alpha_3 \cos\theta_3 - \frac{1}{2}L_3 \omega_3^2 \sin\theta_3$$

Since the slider moves along a straight line, the linear acceleration of its mass center and the linear acceleration of link 4, the slider, are the same vector:

$$a_{xG_4} = a_4$$

$$a_{yG_4} = 0$$

6.5   Four-bar Linkage

To perform an acceleration analysis on a 4-bar linkage, we first write the position equation based on a vector loop and then differentiate it twice with respect to time. The input link is link two, so θ_2, ω_2, and α_2 are known (see Figure 6.6):

$$\overline{L_2} + \overline{L_3} - \overline{L_4} - \overline{L_1} = 0$$

$$L_2\angle\theta_2 + L_3\angle\theta_3 - L_4\angle\theta_4 - L_1\angle\theta_1 = 0$$

$$L_2 e^{i\theta_2} + L_3 e^{i\theta_3} - L_4 e^{i\theta_4} - L_1 e^{i\theta_1} = 0$$

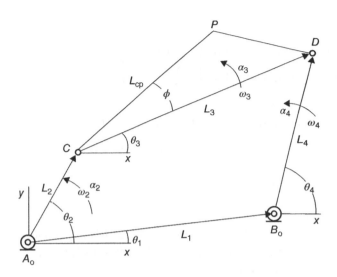

Figure 6.6 Four-bar linkage

Differentiating with respect to time generates the velocity loop equation:

$$L_2 i\omega_2 e^{i\theta_2} + L_3 i\omega_3 e^{i\theta_3} - L_4 i\omega_4 e^{i\theta_4} - L_1 i\omega_1 e^{i\theta_1} = 0$$

Differentiating with respect to time again generates the acceleration loop equation:

$$L_2 i\alpha_2 e^{i\theta_2} - L_2 \omega_2^2 e^{i\theta_2} + L_3 i\alpha_3 e^{i\theta_3} - L_3 \omega_3^2 e^{i\theta_3} - L_4 i\alpha_4 e^{i\theta_4} + L_4 \omega_4^2 e^{i\theta_4} = 0$$

Substituting $e^{i\theta} = \cos\theta + i\sin\theta$ and collecting the real and imaginary terms lead to

$$L_2 i\alpha_2 (\cos\theta_2 + i\sin\theta_2) - L_2 \omega_2^2 (\cos\theta_2 + i\sin\theta_2) + L_3 i\alpha_3 (\cos\theta_3 + i\sin\theta_3)$$

$$- L_3 \omega_3^2 (\cos\theta_3 + i\sin\theta_3) - L_4 i\alpha_4 (\cos\theta_4 + i\sin\theta_4) + L_4 \omega_4^2 (\cos\theta_4 + i\sin\theta_4) = 0$$

real: $-L_2 \alpha_2 \sin\theta_2 - L_2 \omega_2^2 \cos\theta_2 - L_3 \alpha_3 \sin\theta_3 - L_3 \omega_3^2 \cos\theta_3 + L_4 \alpha_4 \sin\theta_4 + L_4 \omega_4^2 \cos\theta_4 = 0$

img: $+L_2 \alpha_2 \cos\theta_2 - L_2 \omega_2^2 \sin\theta_2 + L_3 \alpha_3 \cos\theta_3 - L_3 \omega_3^2 \sin\theta_3 - L_4 \alpha_4 \cos\theta_4 + L_4 \omega_4^2 \sin\theta_4 = 0$

Rewriting in matrix form for the two unknowns α_3 and α_4 shows us that we have two linear equations for the two unknowns:

$$\begin{bmatrix} -L_3 \sin\theta_3 & L_4 \sin\theta_4 \\ L_3 \cos\theta_3 & -L_4 \cos\theta_4 \end{bmatrix} \begin{Bmatrix} \alpha_3 \\ \alpha_4 \end{Bmatrix} = \begin{Bmatrix} L_2 \alpha_2 \sin\theta_2 + L_2 \omega_2^2 \cos\theta_2 + L_3 \omega_3^2 \cos\theta_3 - L_4 \omega_4^2 \cos\theta_4 \\ -L_2 \alpha_2 \cos\theta_2 + L_2 \omega_2^2 \sin\theta_2 + L_3 \omega_3^2 \sin\theta_3 - L_4 \omega_4^2 \sin\theta_4 \end{Bmatrix}$$

or

$$\begin{bmatrix} -L_3\sin\theta_3 & L_4\sin\theta_4 \\ L_3\cos\theta_3 & -L_4\cos\theta_4 \end{bmatrix} \begin{Bmatrix} \alpha_3 \\ \alpha_4 \end{Bmatrix} = \begin{Bmatrix} \text{RHS}_{real} \\ \text{RHS}_{img} \end{Bmatrix}$$

where

$$\text{RHS}_{real} = L_2\alpha_2\sin\theta_2 + L_2\omega_2^2\cos\theta_2 + L_3\omega_3^2\cos\theta_3 - L_4\omega_4^2\cos\theta_4$$

$$\text{RHS}_{img} = -L_2\alpha_2\cos\theta_2 + L_2\omega_2^2\sin\theta_2 + L_3\omega_3^2\sin\theta_3 - L_4\omega_4^2\sin\theta_4$$

Cramer's Rule is the easiest way to obtain the solution for the two unknowns:

$$\alpha_3 = \frac{\begin{vmatrix} \text{RHS}_{real} & L_4\sin\theta_4 \\ \text{RHS}_{img} & -L_4\cos\theta_4 \end{vmatrix}}{\begin{vmatrix} -L_3\sin\theta_3 & L_4\sin\theta_4 \\ L_3\cos\theta_3 & -L_4\cos\theta_4 \end{vmatrix}} = \frac{-\text{RHS}_{real}L_4\cos\theta_4 - \text{RHS}_{img}L_4\sin\theta_4}{L_3L_4(\sin\theta_3\cos\theta_4 - \cos\theta_3\sin\theta_4)}$$

$$\alpha_3 = \frac{-\left(\text{RHS}_{real}\cos\theta_4 + \text{RHS}_{img}\sin\theta_4\right)}{L_3(\sin\theta_3\cos\theta_4 - \cos\theta_3\sin\theta_4)} = \frac{-\left(\text{RHS}_{real}\cos\theta_4 + \text{RHS}_{img}\sin\theta_4\right)}{L_3\sin(\theta_3 - \theta_4)}$$

Now, let's substitute in the expressions for the right-hand side and combine terms:

$$\text{RHS}_{real}\cos\theta_4 = L_2\alpha_2\sin\theta_2\cos\theta_4 + L_2\omega_2^2\cos\theta_2\cos\theta_4 + L_3\omega_3^2\cos\theta_3\cos\theta_4 - L_4\omega_4^2\cos\theta_4\cos\theta_4$$

$$\text{RHS}_{img}\sin\theta_4 = -L_2\alpha_2\cos\theta_2\sin\theta_4 + L_2\omega_2^2\sin\theta_2\sin\theta_4 + L_3\omega_3^2\sin\theta_3\sin\theta_4 - L_4\omega_4^2\sin\theta_4\sin\theta_4$$

collecting terms:

$$L_2\alpha_2(\sin\theta_2\cos\theta_4 - \cos\theta_2\sin\theta_4) + L_2\omega_2^2(\cos\theta_2\cos\theta_4 + \sin\theta_2\sin\theta_4)\ldots$$

$$+ L_3\omega_3^2(\cos\theta_3\cos\theta_4 + \sin\theta_3\sin\theta_4) - L_4\omega_4^2(\cos\theta_4\cos\theta_4 + \sin\theta_4\sin\theta_4)$$

which simplifies to:

$$L_2\alpha_2\sin(\theta_2 - \theta_4) + L_2\omega_2^2\cos(\theta_2 - \theta_4) + L_3\omega_3^2\cos(\theta_3 - \theta_4) - L_4\omega_4^2$$

so that

$$\boxed{\overline{\alpha_3} = \frac{-\left(L_2\alpha_2\sin(\theta_2 - \theta_4) + L_2\omega_2^2\cos(\theta_2 - \theta_4) + L_3\omega_3^2\cos(\theta_3 - \theta_4) - L_4\omega_4^2\right)}{L_3\sin(\theta_3 - \theta_4)}} \quad (+\text{c.c.w.}) \qquad (6.4)$$

$$\alpha_4 = \frac{\begin{vmatrix} -L_3\sin\theta_3 & \text{RHS}_{real} \\ L_3\cos\theta_3 & \text{RHS}_{img} \end{vmatrix}}{\begin{vmatrix} -L_3\sin\theta_3 & L_4\sin\theta_4 \\ L_3\cos\theta_3 & -L_4\cos\theta_4 \end{vmatrix}} = \frac{-L_3\sin\theta_3\text{RHS}_{img} - L_3\cos\theta_3\text{RHS}_{real}}{L_3L_4(\sin\theta_3\cos\theta_4 - \cos\theta_3\sin\theta_4)}$$

$$\alpha_4 = \frac{-\left(\text{RHS}_{real}\cos\theta_3 + \text{RHS}_{img}\sin\theta_3\right)}{L_4(\sin\theta_3\cos\theta_4 - \cos\theta_3\sin\theta_4)}$$

Now, let's substitute in the expressions for the right-hand side and combine terms:

$$\text{RHS}_{\text{real}}\cos\theta_3 = L_2\alpha_2\sin\theta_2\cos\theta_3 + L_2\omega_2^2\cos\theta_2\cos\theta_3 + L_3\omega_3^2\cos\theta_3\cos\theta_3 - L_4\omega_4^2\cos\theta_4\cos\theta_3$$

$$\text{RHS}_{\text{img}}\sin\theta_3 = -L_2\alpha_2\cos\theta_2\sin\theta_3 + L_2\omega_2^2\sin\theta_2\sin\theta_3 + L_3\omega_3^2\sin\theta_3\sin\theta_3 - L_4\omega_4^2\sin\theta_4\sin\theta_3$$

collecting terms:

$$L_2\alpha_2(\sin\theta_2\cos\theta_3 - \cos\theta_2\sin\theta_3) + L_2\omega_2^2(\cos\theta_2\cos\theta_3 + \sin\theta_2\sin\theta_3)\dots$$

$$+ L_3\omega_3^2(\cos\theta_3\cos\theta_3 + \sin\theta_3\sin\theta_3) - L_4\omega_4^2(\cos\theta_4\cos\theta_3 + \sin\theta_4\sin\theta_3)$$

which simplifies to:

$$L_2\alpha_2\sin(\theta_2-\theta_3) + L_2\omega_2^2\cos(\theta_2-\theta_3) + L_3\omega_3^2 - L_4\omega_4^2\cos(\theta_3-\theta_4)$$

so that

$$\boxed{\overline{\alpha_4} = \frac{L_2\alpha_2\sin(\theta_2-\theta_3) + L_2\omega_2^2\cos(\theta_2-\theta_3) + L_3\omega_3^2 - L_4\omega_4^2\cos(\theta_4-\theta_3)}{L_4\sin(\theta_4-\theta_3)}} \quad (+\text{c.c.w.}) \qquad (6.5)$$

From the position analysis for a 4-bar linkage, two nonlinear equations are used to solve for the angular position of links 3 and 4. This can be done in a number of ways (see Chapter 2):

$$L_3\sin\theta_3 - L_4\sin\theta_4 = L_1\sin\theta_1 - L_2\sin\theta_2$$

$$L_3\cos\theta_3 - L_4\cos\theta_4 = L_1\cos\theta_1 - L_2\cos\theta_2$$

From the velocity analysis for a 4-bar linkage with counter clockwise being positive,

$$\boxed{\begin{aligned}\overline{\omega_3} &= \frac{-\omega_2 L_2\sin(\theta_2-\theta_4)}{L_3\sin(\theta_3-\theta_4)} \\ \overline{\omega_4} &= \frac{\omega_2 L_2\sin(\theta_2-\theta_3)}{L_4\sin(\theta_4-\theta_3)}\end{aligned}} \quad (+\text{c.c.w.}) \qquad\begin{aligned}(5.1)\\[1.5em](5.2)\end{aligned}$$

Example 6.2   Acceleration analysis of a 4-bar linkage

Problem: A 4-bar linkage (Figure 6.7) has link lengths of $L_1 = 7.80$ in., $L_2 = 2.90$ in., $L_3 = 7.25$ in., and $L_4 = 3.50$ in. Bearings A_o and B_o are located on a horizontal line. Using the current position, $\theta_2 = 120°$, the position analysis has determined that $\theta_3 = -33.2°$ and $\theta_4 = 205°$. Currently, link 2 is traveling at 5.50 rad/s counterclockwise and accelerating at a rate of 2.2 rad/s^2. What are the angular acceleration vectors for links 3 and 4?

Solution:

1. Verify the position vectors given:

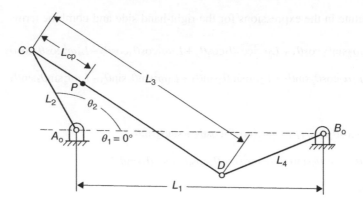

Figure 6.7 Four-bar linkage acceleration analysis

$$C_1 = \frac{L_1^2 + L_2^2 + L_4^2 - L_3^2}{2L_4} - \frac{L_1 L_2 \cos(\theta_1 - \theta_2)}{L_4} = 7.365$$

$$C_2 = L_1 \cos\theta_1 - L_2 \cos\theta_2 = 9.25$$

$$C_3 = L_1 \sin\theta_1 - L_2 \sin\theta_2 = -2.511$$

From the figure, note that θ_4 is in the third quadrant and θ_3 is in the fourth quadrant:

$$\theta_4 = 2\tan^{-1}\left(\frac{-C_3 + \sqrt{C_3^2 - C_1^2 + C_2^2}}{(C_1 - C_2)}\right) = -155.4° = 204.6°$$

$$\theta_3 = \tan^{-1}\left(\frac{L_1 \sin\theta_1 + L_4 \sin\theta_4 - L_2 \sin\theta_2}{L_1 \cos\theta_1 + L_4 \cos\theta_4 - L_2 \cos\theta_2}\right) = -33.2°$$

2. Calculate the velocity vectors:

$$\omega_3 = -\omega_2 \frac{L_2 \sin(\theta_2 - \theta_4)}{L_3 \sin(\theta_3 - \theta_4)} = \frac{-5.5(2.90)\sin(120° - 204.6°)}{7.25 \sin(-33.2° - 204.6°)} = \frac{15.88}{6.201} = 2.589 \text{ rad/s}$$

$$\omega_4 = \frac{\omega_2 L_2 \sin(\theta_2 - \theta_3)}{L_4 \sin(\theta_4 - \theta_3)} = \frac{5.5(2.90)\sin(120° - (-33.2°))}{3.50 \sin(204.6° - (-33.2°))} = \frac{7.192}{-2.962} = -2.428 \text{ rad/s}$$

3. Calculate the acceleration vectors:

$$\overline{\alpha_3} = \frac{-\left(L_2 \alpha_2 \sin(\theta_2 - \theta_4) + L_2 \omega_2^2 \cos(\theta_2 - \theta_4) + L_3 \omega_3^2 \cos(\theta_3 - \theta_4) - L_4 \omega_4^2\right)}{L_3 \sin(\theta_3 - \theta_4)}$$

$$\overline{\alpha_3} = \frac{-(2.90)(2.2)\sin(120° - 204.6°) + 2.90(5.5)^2 \cos(120° - 204.6°) + 7.25(2.589)^2 \cos(-33.2° - 204.6°) - 3.50(-2.428)^2}{7.25 \sin(-33.2° - 204.6°)}$$

$$\overline{\alpha_3} = 7.282 \text{ rad/s}^2$$

$$\overline{\alpha_4} = \frac{L_2\alpha_2\sin(\theta_2-\theta_3)+L_2\omega_2^2\cos(\theta_2-\theta_3)+L_3\omega_3^2-L_4\omega_4^2\cos(\theta_4-\theta_3)}{L_4\sin(\theta_4-\theta_3)}$$

$$\overline{\alpha_4} = \frac{2.90(2.2)\sin(120°-(-33.2°))+2.90(5.5)^2\cos(120°-(-33.2°))+7.25(2.589)^2-3.50(-2.428)^2\cos(204.6°-(-33.2°))}{3.50\sin(204.6°-(-33.2°))}$$

$$\overline{\alpha_4} = 5.337 \text{ rad/s}^2$$

Answers: $\boxed{\bar{\alpha}_3 = 7.28 \text{ rad/s}^2 \text{ c.c.w.}}$ and $\boxed{\bar{\alpha}_4 = 5.34 \text{ rad/s}^2 \text{ c.c.w.}}$

6.6 Acceleration of Mass Centers for 4-bar Linkage

To perform a dynamic force analysis, the accelerations of the mass centers of all moving links must be determined. If we assume the links are made from homogeneous material, then their mass centers will be located at their geometric centers (see Figure 6.8) where G_2, G_3, and G_4 represent the mass centers of links 2, 3, and 4.

Mass center G_2 is located at the midpoint of link 2; thus,

$$\overline{L_{G_2}} = \frac{1}{2}\overline{L_2} = \frac{1}{2}L_2e^{i\theta_2}$$

Differentiating twice with respect to time leads to

$$\overline{a_{G_2}} = \frac{1}{2}L_2i\alpha_2e^{i\theta_2} + \frac{1}{2}L_2i^2\omega_2^2e^{i\theta_2}$$

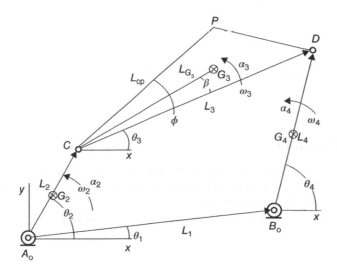

Figure 6.8 Mass centers for 4-bar linkage

$$a_{xG_2} = -\frac{1}{2}L_2\alpha_2 \sin\theta_2 - \frac{1}{2}L_2\omega_2^2 \cos\theta_2$$

$$a_{yG_2} = \frac{1}{2}L_2\alpha_2 \cos\theta_2 - \frac{1}{2}L_2\omega_2^2 \sin\theta_2$$

Mass center G_3 is located at the midpoint of link 3; thus,

$$\overline{L_{G_3}} = \overline{L_2} + \overline{L_{G_3}} = L_2 e^{i\theta_2} + L_{G_3}e^{i(\theta_3 + \beta)}$$

Differentiating twice with respect to time leads to

$$\overline{a_{G_3}} = L_2 i\alpha_2 e^{i\theta_2} + L_2 i^2\omega_2^2 e^{i\theta_2} + L_{G_3}i\alpha_3 e^{i(\theta_3+\beta)} + L_{G_3}i^2\omega_3^2 e^{i(\theta_3+\beta)}$$

$$a_{xG_3} = -L_2\alpha_2 \sin\theta_2 - L_2\omega_2^2\cos\theta_2 - L_{G_3}\alpha_3\sin(\theta_3+\beta) - L_{G_3}\omega_3^2\cos(\theta_3+\beta)$$

$$a_{yG_3} = L_2\alpha_2 \cos\theta_2 - L_2\omega_2^2\sin\theta_2 + L_{G_3}\alpha_3\cos(\theta_3+\beta) - L_{G_3}\omega_3^2\sin(\theta_3+\beta)$$

Since link 4 is connected to ground, its acceleration vector looks identical to link 2's acceleration vector except for the subscript:

$$\overline{L_{G_4}} = \overline{L_1} + \frac{1}{2}\overline{L_4} = L_1 e^{i\theta_1} + \frac{1}{2}L_4 e^{i\theta_4}$$

Differentiating twice with respect to time leads to

$$\overline{a_{G_4}} = \frac{1}{2}L_4 i\alpha_4 e^{i\theta_4} + \frac{1}{2}L_4 i^2\omega_4^2 e^{i\theta_4}$$

$$a_{xG_4} = -\frac{1}{2}L_4\alpha_4 \sin\theta_4 - \frac{1}{2}L_4\omega_4^2\cos\theta_4$$

$$a_{yG_4} = \frac{1}{2}L_4\alpha_4 \cos\theta_4 - \frac{1}{2}L_4\omega_4^2\sin\theta_4$$

6.7   Coriolis Acceleration

Coriolis acceleration shows up when an object is moving along a path and the path is rotating about a point at this instant. As seen in Figure 6.9, link 3 slides along link 4, while link 4 rotates about bearing B_o; thus, Coriolis acceleration is present.

The vector loop equation for this linkage contains only L_1, L_2, and L_3:

$$\overline{L_1} + \overline{L_2} = \overline{L_3}$$

$$L_1 e^{i\theta_1} + L_2 e^{i\theta_2} = L_3 e^{i\theta_3}$$

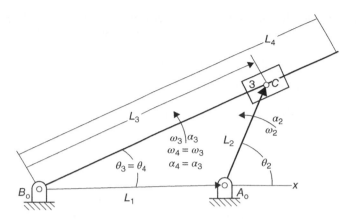

Figure 6.9   Inverted slider–crank

Solving for L_3 and θ_3,

$$\theta_3 = \tan^{-1}\left(\frac{L_2 \sin\theta_2}{L_1 + L_2 \cos\theta_2}\right) \tag{6.6}$$

$$L_3 = L_1 + L_2 \ \{\text{when } \theta_2 = 0°\} \tag{6.7}$$

$$L_3 = L_1 - L_2 \ \{\text{when } \theta_2 = 180°\}$$

$$L_3 = \frac{L_2 \sin\theta_2}{\sin\theta_3} \ \{\text{when } \theta_2 \neq 0° \text{ or } 180°\}$$

Differentiate the position equation with respect to time and solve for $v_{3/4}$ and ω_3:

$$L_2 i\omega_2 e^{i\theta_2} = v_{3/4}e^{i\theta_3} + L_3 i\omega_3 e^{i\theta_3}$$

$$v_{3/4} = L_2\omega_2 \sin(\theta_3 - \theta_2)\angle\theta_3 \tag{5.5}$$

$$\omega_3 = \frac{L_2}{L_3}\omega_2 \cos(\theta_3 - \theta_2) \tag{5.6}$$

$(+\text{c.c.w.})$

Differentiating the velocity equation with respect to time leads to

$$L_2 i\alpha_2 e^{i\theta_2} + L_2 i^2 \omega_2^2 e^{i\theta_2} = \left(a_{3/4}e^{i\theta_3} + v_{3/4}i\omega_3 e^{i\theta_3}\right) + \left(v_{3/4}i\omega_3 e^{i\theta_3} + L_3 i\alpha_3 e^{i\theta_3} + L_3 i^2 \omega_3^2 e^{i\theta_3}\right)$$

collecting the two similar terms

$$L_2 i\alpha_2 e^{i\theta_2} + L_2 i^2 \omega_2^2 e^{i\theta_2} = a_{3/4}e^{i\theta_3} + 2v_{3/4}i\omega_3 e^{i\theta_3} + L_3 i\alpha_3 e^{i\theta_3} + L_3 i^2 \omega_3^2 e^{i\theta_3}$$

where
$2v_{3/4}i\omega_3 e^{i\theta_3}$ is the Coriolis acceleration term.

Note that for the Coriolis acceleration term to be nonzero, link 3 needs to have a linear velocity along link 4, and link 4 needs to have an angular velocity about the fixed bearing B_o:

$$L_2\alpha_2\cos(\theta_2+90°)+L_2\omega_2^2\cos(\theta_2+180°)=a_{3/4}\cos\theta_3+2v_{3/4}\omega_3\cos(\theta_3+90°)\ldots$$
$$+L_3\alpha_3\cos(\theta_3+90°)+L_3\omega_3^2\cos(\theta_3+180°)$$

$$L_2\alpha_2\sin(\theta_2+90°)+L_2\omega_2^2\sin(\theta_2+180°)=a_{3/4}\sin\theta_3+2v_{3/4}\omega_3\sin(\theta_3+90°)\ldots$$
$$+L_3\alpha_3\sin(\theta_3+90°)+L_3\omega_3^2\sin(\theta_3+180°)$$

Simplifying,

$$-L_2\alpha_2\sin\theta_2-L_2\omega_2^2\cos\theta_2=a_{3/4}\cos\theta_3-2v_{3/4}\omega_3\sin\theta_3-L_3\alpha_3\sin\theta_3-L_3\omega_3^2\cos\theta_3$$
$$+L_2\alpha_2\cos\theta_2-L_2\omega_2^2\sin\theta_2=a_{3/4}\sin\theta_3+2v_{3/4}\omega_3\cos\theta_3+L_3\alpha_3\cos\theta_3-L_3\omega_3^2\sin\theta_3$$

Rewriting the two equations in matrix form,

$$\begin{bmatrix}\cos\theta_3 & -L_3\sin\theta_3\\ \sin\theta_3 & L_3\cos\theta_3\end{bmatrix}\begin{Bmatrix}a_{3/4}\\ a_{3/4}\end{Bmatrix}=\begin{Bmatrix}-L_2\alpha_2\sin\theta_2-L_2\omega_2^2\cos\theta_2+2v_{3/4}\omega_3\sin\theta_3+L_3\omega_3^2\cos\theta_3\\ +L_2\alpha_2\cos\theta_2-L_2\omega_2^2\sin\theta_2-2v_{3/4}\omega_3\cos\theta_3+L_3\omega_3^2\sin\theta_3\end{Bmatrix}$$

or

$$\begin{bmatrix}\cos\theta_3 & -L_3\sin\theta_3\\ \sin\theta_3 & L_3\cos\theta_3\end{bmatrix}\begin{Bmatrix}a_{3/4}\\ a_{3/4}\end{Bmatrix}=\begin{Bmatrix}\text{RHS}_{\text{real}}\\ \text{RHS}_{\text{img}}\end{Bmatrix}$$

where

$$\text{RHS}_{\text{real}}=-L_2\alpha_2\sin\theta_2-L_2\omega_2^2\cos\theta_2+2v_{3/4}\omega_3\sin\theta_3+L_3\omega_3^2\cos\theta_3$$
$$\text{RHS}_{\text{img}}=+L_2\alpha_2\cos\theta_2-L_2\omega_2^2\sin\theta_2-2v_{3/4}\omega_3\cos\theta_3+L_3\omega_3^2\sin\theta_3$$

Cramer's Rule is the easiest way to obtain the solution for the two unknowns:

$$a_{3/4}=\frac{\begin{vmatrix}\text{RHS}_{\text{real}} & -L_3\sin\theta_3\\ \text{RHS}_{\text{img}} & L_3\cos\theta_3\end{vmatrix}}{\begin{vmatrix}\cos\theta_3 & -L_3\sin\theta_3\\ \sin\theta_3 & L_3\cos\theta_3\end{vmatrix}}=\frac{\text{RHS}_{\text{real}}L_3\cos\theta_3+\text{RHS}_{\text{img}}L_3\sin\theta_3}{L_3\left(\cos^2\theta_3+\sin^2\theta_3\right)}$$

$$a_{3/4}=\text{RHS}_{\text{real}}\cos\theta_3+RHS_{\text{img}}\sin\theta_3$$

$$a_{3/4}=-L_2\alpha_2\sin\theta_2\cos\theta_3-L_2\omega_2^2\cos\theta_2\cos\theta_3+2v_{3/4}\omega_3\sin\theta_3\cos\theta_3+L_3\omega_3^2\cos\theta_3\cos\theta_3\ldots$$

$$+L_2\alpha_2\cos\theta_2\sin\theta_3-L_2\omega_2^2\sin\theta_2\sin\theta_3-2v_{3/4}\omega_3\cos\theta_3\sin\theta_3+L_3\omega_3^2\sin\theta_3\sin\theta_3$$

$$\boxed{a_{3/4}\angle\theta_3=-L_2\alpha_2\sin(\theta_2-\theta_3)-L_2\omega_2^2\cos(\theta_2-\theta_3)+L_3\omega_3^2} \qquad (6.8)$$

$$\alpha_3 = \frac{\begin{vmatrix} \cos\theta_3 & RHS_{real} \\ \sin\theta_3 & RHS_{img} \end{vmatrix}}{\begin{vmatrix} \cos\theta_3 & -L_3\sin\theta_3 \\ \sin\theta_3 & L_3\cos\theta_3 \end{vmatrix}} = \frac{RHS_{img}\cos\theta_3 - RHS_{real}\sin\theta_3}{L_3(\cos^2\theta_3 + \sin^2\theta_3)}$$

$$\alpha_3 = (L_2\alpha_2\cos\theta_2\cos\theta_3 - L_2\omega_2^2\sin\theta_2\cos\theta_3 - 2v_{3/4}\omega_3\cos\theta_3\cos\theta_3 + L_3\omega_3^2\sin\theta_3\cos\theta_3 \ldots$$

$$+ L_2\alpha_2\sin\theta_2\sin\theta_3 + L_2\omega_2^2\cos\theta_2\sin\theta_3 - 2v_{3/4}\omega_3\sin\theta_3\sin\theta_3 - L_3\omega_3^2\cos\theta_3\sin\theta_3)/L_3$$

$$\boxed{\overline{\alpha_3} = \frac{L_2\alpha_2}{L_3}\cos(\theta_2 - \theta_3) - \frac{L_2\omega_2^2}{L_3}\sin(\theta_2 - \theta_3) - \frac{2v_{3/4}\omega_3}{L_3}} \quad (+\text{c.c.w. and }+v_{3/4}\angle\theta_3) \qquad (6.9)$$

The acceleration vectors for other linkages can be found in a similar manner using the vector loop approach and differentiating twice with respect to time.

Example 6.3 Acceleration analysis for an inverted slider–crank mechanism

Problem: Given the inverted slider–crank mechanism shown in Figure 6.10, find the angular acceleration vector for links 3 and 4 and the linear acceleration vector for link 3 relative to link 4. Link 1 (from A_o to B_o) is 6.25 in. and at an angle of 172°. Link 2 is 1.75 in. long and at an angle of 46°. Link 3 is 5.70 in. from pin C to the top of the piston. Currently, the top of the piston, link 4, is 1.72 in. from bearing B_o and at an angle of 3.01°. Link 2 is traveling at a constant 6 rad/s clockwise.

Solution:

1. Verify the position data given:

$$\overline{L_1} + \overline{L_4} = \overline{L_2} + \overline{L_3}$$

$$L_1\sin\theta_1 + L_4\sin\theta_4 = L_2\sin\theta_2 + L_3\sin\theta_3$$

$$L_1\cos\theta_1 + L_4\cos\theta_4 = L_2\cos\theta_2 + L_3\cos\theta_3$$

Or a better way to solve for L_4 is using the cosine law:

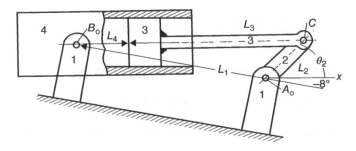

Figure 6.10 Inverted slider–crank acceleration analysis

$$(L_3 + L_4)^2 = L_1^2 + L_2^2 - 2L_1 L_2 \cos(\theta_1 - \theta_2)$$

$$L_4 = \sqrt{6.25^2 + 1.75^2 - 2(6.25)(1.75)\cos(172° - 46°)} - 5.70$$

$$L_4 = 1.715 \text{ in.}$$

Solving for θ_4 using the y-component equation and noting that $\theta_3 = \theta_4 + 180°$,

$$\theta_4 = \sin^{-1}\left(\frac{L_2 \sin\theta_2 - L_1 \sin\theta_1}{L_3 + L_4}\right)$$

$$\theta_4 = \sin^{-1}\left(\frac{1.75\sin 46° - 6.25 \sin 172°}{5.70 + 1.715}\right) = \sin^{-1}(0.0525) = 3.007°$$

2. Calculate the velocity vectors and noting that $\omega_3 = \omega_4$,

$$V_{3/4}\angle\theta_4 + L_4\omega_4\angle(\theta_4 + 90°) = L_2\omega_2\angle(\theta_2 + 90°) + L_3\omega_3\angle(\theta_3 + 90°)$$

or

$$V_{3/4}\angle\theta_4 + L_4\omega_4\angle(\theta_4 + 90°) = L_2\omega_2\angle(\theta_2 + 90°) + L_3\omega_4\angle(\theta_4 + 180° + 90°)$$

$$\begin{bmatrix} \sin\theta_3 & (L_3 + L_4)\cos\theta_3 \\ \cos\theta_3 & -(L_3 + L_4)\sin\theta_3 \end{bmatrix} \begin{Bmatrix} V_{3/4} \\ \omega_4 \end{Bmatrix} = \begin{Bmatrix} L_2\omega_2 \cos\theta_2 \\ -L_2\omega_2 \sin\theta_2 \end{Bmatrix}$$

Noting that this is the same as the development of the inverted slider–crank shown in Figure 5.30, we can conclude that the equations in Chapter 5 apply if we let the sum of L_3 and L_4 in this example be equal to the length of L_3 in Chapter 5:

$$\overline{V_{3/4}} = \omega_2 L_2 \sin(\theta_4 - \theta_2)\angle\theta_4$$

$$\overline{V_{3/4}} = -6.00(1.75)\sin(3° - 46°) = 7.161 \text{ in./s}\angle 3°$$

$$\overline{\omega_4} = \frac{\omega_2 L_2 \cos(\theta_4 - \theta_2)}{(L_3 + L_4)}$$

$$\overline{\omega_4} = \left(\frac{-6.00(1.75)\cos(3° - 46°)}{(5.70 + 1.715)}\right) = -1.036 \text{ rad/s}$$

3. Calculate the acceleration vectors using the similarities noted above:

$$a_{3/4}\angle\theta_4 = -L_2\alpha_2 \sin(\theta_2 - \theta_4) - L_2\omega_2^2 \cos(\theta_2 - \theta_4) + (L_3 + L_4)\omega_3^2$$

$$a_{3/4}\angle\theta_4 = -1.75(0)\sin(46° - 3°) - 1.75(-6.00)^2 \cos(46° - 3°) + (5.70 + 1.715)(-1.036)^2$$

$$a_{3/4} = -38.12 \text{ in./s}^2 \angle 3° = 38.12 \text{ in./s}^2 \angle 183°$$

$$\overline{\alpha_4} = \frac{L_2\alpha_2}{(L_3+L_4)}\cos(\theta_2-\theta_4) - \frac{L_2\omega_2^2}{(L_3+L_4)}\sin(\theta_2-\theta_4) - \frac{2v_{3/4}\omega_4}{(L_3+L_4)}$$

$$\overline{\alpha_4} = \frac{1.75(0)}{(5.70+1.715)}\cos(46°-3°) - \frac{1.75(-6.00)^2}{(5.70+1.715)}\sin(46°-3°) - \frac{2(7.161)(-1.036)}{(5.70+1.715)}$$

$$\overline{\alpha_4} = -3.793$$

$$\overline{\alpha_4} = 3.793 \text{ rad/s}^2 \text{ c.w.}$$

Answers:

$$\boxed{\overline{\alpha_4} = \overline{\alpha_3} = 3.79 \text{ rad/s}^2 \text{ c.w.}}, \text{ and } \boxed{\overline{a_{3/4}} = 38.1 \text{ in./s}^2 \angle 183°}.$$

Problems

6.1   Given (Figure 6.11): $L_1 = 185$ mm, $L_2 = 50.0$ mm, $L_3 = 180$ mm, $L_4 = 115$ mm, $\theta_1 = 3°$, and $\theta_2 = 152°$. A position analysis has determined that the angular positions of links 3 and 4 are $\theta_3 = 26.2°$ and $\theta_4 = 125.9°$. Assuming that link 2 is traveling at a constant 12 rad/s counterclockwise, determine the angular acceleration vectors for links 3 and 4. A velocity analysis has determined that the angular velocity vectors for links 3 and 4 are $\omega_3 = 1.490$ rad/s counter clockwise and $\omega_4 = 4.292$ rad/s counter clockwise. (Answers: $\alpha_3 = 24.1$ rad/s² counter clockwise, $\alpha_4 = 30.5$ rad/s² clockwise).

6.2   Given (Figure 6.11): $L_1 = 185$ mm, $L_2 = 50.0$ mm, $L_3 = 180$ mm, $L_4 = 115$ mm, $\theta_1 = 3°$, and $\theta_2 = 152°$. A position analysis has determined that the angular positions of links 3 and 4 are $\theta_3 = 26.2°$ and $\theta_4 = 125.9°$. Assuming that link 2 is traveling at 12 rad/s counterclockwise and increasing speed at 2 rad/s², determine the angular acceleration vectors for links 3 and 4. A velocity analysis has determined that the angular velocity vectors for links 3 and 4 are $\omega_3 = 1.490$ rad/s counter clockwise and $\omega_4 = 4.292$ rad/s counter clockwise.

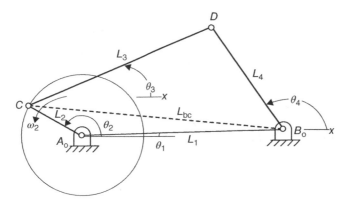

Figure 6.11   Problems 6.1 and 6.2

6.3 Given (Figure 6.12): $L_1 = 8.50$ in., $L_2 = 3.00$ in., $L_3 = 5.00$ in., $L_4 = 4.50$ in., $\theta_1 = 0°$, and $\theta_2 = 60°$. A position analysis has determined that the angular positions of links 3 and 4 are $\theta_3 = 15.5°$ and $\theta_4 = 119.0°$. Assuming that link 2 is traveling at a constant 10 rad/s clockwise (c.w.), determine the angular acceleration vectors for links 3 and 4. A velocity analysis has determined that the angular velocity vectors for links 3 and 4 are $\omega_3 = 5.289$ rad/s counter clockwise and $\omega_4 = 4.804$ rad/s clockwise.

6.4 Given (Figure 6.12): $L_1 = 8.50$ in., $L_2 = 3.00$ in., $L_3 = 5.00$ in., $L_4 = 4.50$ in., $\theta_1 = 0°$, and $\theta_2 = 60°$. A position analysis has determined that the angular positions of links 3 and 4 are $\theta_3 = 15.5°$ and $\theta_4 = 119.0°$. Assuming that link 2 is traveling at 10 rad/s clockwise and accelerating at 2 rad/s^2, determine the angular acceleration vectors for links 3 and 4. A velocity analysis has determined that the angular velocity vectors for links 3 and 4 are $\omega_3 = 5.289$ rad/s counter clockwise and $\omega_4 = 4.804$ rad/s clockwise.

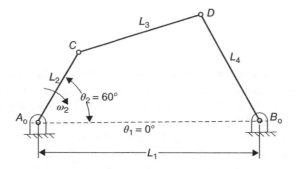

Figure 6.12 Problems 6.3 and 6.4

6.5 Given (Figure 6.13): $L_1 = 175$ mm, $L_2 = 65.0$ mm, $L_3 = 120$ mm, $L_4 = 105$ mm, $\theta_1 = 20°$, and $\theta_2 = 312°$. A position analysis has determined that the angular positions of links 3 and 4 are $\theta_3 = 82.1°$ and $\theta_4 = 174.1°$. Assuming that link 2 is traveling at a constant 15 rad/s counterclockwise, determine the angular acceleration vectors for links 3 and 4. A velocity analysis has determined that the angular velocity vectors for links 3 and 4 are $\omega_3 = 5.456$ rad/s counter clockwise and $\omega_4 = 7.108$ rad/s clockwise.

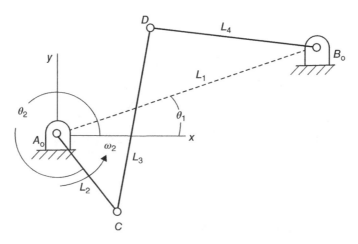

Figure 6.13 Problem 6.5

6.6 Given (Figure 6.14): $L_1 = 175$ mm, $L_2 = 65.0$ mm, $L_3 = 170$ mm, $L_4 = 75.0$ mm $\theta_1 = 335°$, and $\theta_2 = 98°$. Assume, $L_{cp} = 120$ mm, $L_{pd} = 70$ mm. A position analysis has determined that the angular positions of links 3 and 4 are $\theta_3 = 303.1°$ and $\theta_4 = 183.2°$. Assuming that link 2 is traveling at a constant 15 rad/s clockwise, determine the angular acceleration vectors for links 3 and 4. Also, determine the linear acceleration vector for point P. A velocity analysis has determined that the angular velocity vectors for links 3 and 4 are $\omega_3 = 6.593$ rad/s clockwise and $\omega_4 = 6.355$ rad/s counter clockwise.

6.7 Given (Figure 6.14): $L_1 = 175$ mm, $L_2 = 65.0$ mm, $L_3 = 170$ mm, $L_4 = 75.0$ mm, $\theta_1 = 335°$, and $\theta_2 = 98°$. Assume $L_{cp} = 120$ mm, $L_{pd} = 70$ mm. A position analysis has determined that the angular positions of links 3 and 4 are $\theta_3 = 303.1°$ and $\theta_4 = 183.2°$. Assuming that link 2 is traveling at 15 rad/s clockwise and slowing at a rate of 2 rad/s^2, determine the angular acceleration vectors for links 3 and 4. Also, determine the linear acceleration vector for point P. A velocity analysis has determined that the angular velocity vectors for links 3 and 4 are $\omega_3 = 6.593$rad/s clockwise and $\omega_4 = 6.355$ rad/s counter clockwise.

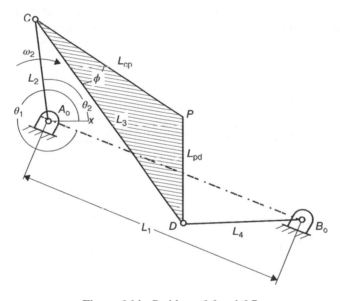

Figure 6.14 Problems 6.6 and 6.7

6.8   Given (Figure 6.15): $L_1 = 3.50$ in., $L_2 = 3.25$ in., $L_3 = 6.50$ in., $\theta_1 = 90°$, $\theta_4 = 0°$, and currently $\theta_2 = 110°$. A position analysis has determined that the positions of links 3 and 4 are $\theta_3 = 3.9°$ and $L_4 = 5.37$ in. Assuming that link 2 is traveling at a constant 4 rad/s counterclockwise, determine the angular acceleration vector for link 3 and the linear acceleration vector for the slider, link 4. A velocity analysis has determined that the angular velocity vector for links 3 is $\omega_3 = 0.6857$ rad/s counter clockwise and the linear velocity vector is $V_4 = 12.52$ in/s$\leftarrow$ (to the left).

6.9   Given (Figure 6.15): $L_1 = 3.50$ in., $L_2 = 3.25$ in., $L_3 = 6.50$ in., $\theta_1 = 90°$, $\theta_4 = 0°$, and currently $\theta_2 = 110°$. A position analysis has determined that the positions of links 3 and 4 are $\theta_3 = 3.9°$ and $L_4 = 5.37$ in. Assuming that link 2 is traveling at 4 rad/s counterclockwise and accelerating at a rate of 4 rad/s^2, determine the angular acceleration vector for link 3 and the linear acceleration vector for the slider, link 4. A velocity analysis has determined that the angular velocity vector for link 3 is $\omega_3 = 0.6857$ rad/s counter clockwise and the linear velocity vector is $V_4 = 12.52$ in/s$\leftarrow$ (to the left).

6.10   Given (Figure 6.15): $L_1 = 90.0$ mm, $L_2 = 85.0$ mm, $L_3 = 170$ mm, $\theta_1 = 90°$, $\theta_4 = 0°$, and $\theta_2 = 110°$. A position analysis has determined that the positions of links 3 and 4 are $\theta_3 = 3.4°$ and $L_4 = 140.6$ mm. Assuming that link 2 is traveling at a constant 14 rad/s counterclockwise, determine the angular acceleration vector for link 3 and the linear acceleration vector for the slider, link 4. A velocity analysis has determined that the angular velocity vector for link 3 is $\omega_3 = 2.398$ rad/s counter clockwise and the linear velocity vector is $V_4 = 1143$ mm/s$\leftarrow$ (to the left).

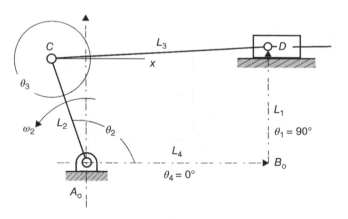

Figure 6.15   Problems 6.8 through 6.10

6.11   Given (Figure 6.16): $L_2 = 56.0$ mm, $L_3 = 130$ mm, $\theta_4 = 90°$, and $\theta_2 = 32°$. A position
       analysis has determined that the positions of links 3 and 4 are $\theta_3 = 111.4°$ and L_4
       $= 150.7$ mm. Assuming that link 2 is traveling at a constant 14 rad/s clockwise,
       determine   the   angular   acceleration   vector   for   link   3   and   the   linear
       acceleration vector for the slider, link 4. A velocity analysis has determined that the
       angular velocity vector for link 3 is $\omega_3 = 3.433$ rad/s counter clockwise and the linear
       velocity vector is $V_4 = 827.9$ mm/s↓ (down).
6.12   Given (Figure 6.16): $L_2 = 56.0$ mm, $L_3 = 130$ mm, $\theta_4 = 90°$, and $\theta_2 = 32°$. A position
       analysis has determined that the positions of links 3 and 4 are $\theta_3 = 111.4°$ and L_4
       $= 150.7$ mm. Assuming that link 2 is traveling at 14 rad/s clockwise and slowing at
       a rate of 4 rad/s^2, determine the angular acceleration vector for link 3 and the linear
       acceleration vector for the slider, link 4. A velocity analysis has determined that the
       angular velocity vector for link 3 is $\omega_3 = 3.433$ rad/s counter clockwise and the linear
       velocity vector is $V_4 = 827.9$ mm/s↓ (down).

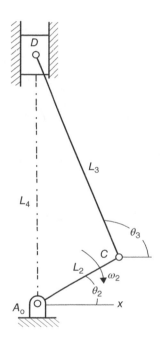

Figure 6.16   Problems 6.11 and 6.12

6.13   Given (Figure 6.17): $L_1 = 5.60$ in., $L_2 = 3.25$ in., $L_4 = 10.0$ in., $\theta_1 = 0$, $\theta_4 = \theta_3$, and currently $\theta_2 = 66°$. A position analysis has determined that the positions of links 3 and 4 are $\theta_4 = 23.2°$ and $L_3 = 7.53$ in. Assuming that link 2 is traveling at a constant 4 rad/s counterclockwise, determine the angular acceleration vector for link 4 and the linear acceleration vector of the slider, link 3, relative to link 4. A velocity analysis has determined that the angular velocity vectors for link 4 and the linear velocity vector for link 3 relative to link 4 are $\omega_4 = 1.267$ rad/s counter clockwise and $V_{3/4} = 8.830$ in./s at 203.2°.

6.14   Given (Figure 6.17): $L_1 = 140$ mm, $L_2 = 85.0$ mm, $L_4 = 250$ mm, $\theta_1 = 0°$, $\theta_4 = \theta_3$, and currently $\theta_2 = 66°$. A position analysis has determined that the positions of links 3 and 4 are $\theta_4 = 24.0°$ and $L_3 = 191.1$ mm. Assuming that link 2 is traveling at a constant 14 rad/s counterclockwise, determine the angular acceleration vector for link 4 and the linear acceleration vector of the slider, link 3, relative to link 4. A velocity analysis has determined that the angular velocity vectors for link 4 and the linear velocity vector for link 3 relative to link 4 are $\omega_4 = 4.627$ rad/s counter clockwise and $V_{3/4} = 796.6$ mm/s at 204°.

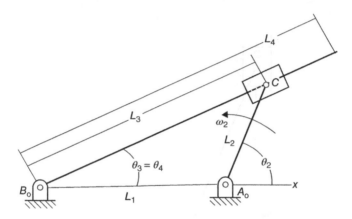

Figure 6.17   Problems 6.13 and 6.14

6.15   Given (Figure 6.18): $L_1 = 75.0$ mm, $L_2 = 70.0$ mm, $L_3 = 100$ mm, $L_4 = 80.0$ mm, $L_5 = 120$ mm, $L_6 = 100$ mm, $L_7 = 115$ mm, $\theta_1 = \theta_7 = 0°$, and currently $\theta_2 = 95°$ and currently rotating at a constant 9.00 rad/s counter clockwise. Also, $L_{cp} = 65.0$ mm, $L_{pd} = 60.0$ mm, $L_{P1P2} = 85.0$ mm, and $L_{P2F} = 60.0$ mm. A positional analysis has determined that the angular positions of links 3, 4, 5, and 6 are $\theta_3 = 4.63°$, $\theta_4 = 76.6°$, $\theta_5 = -7.23°$, and $\theta_6 = 105.7°$. Similarly, the current locations of point P_1 and point P_2 are $P_1 = (43.8, 111)$ mm and $P_2 = (123, 141)$ mm. Also, a velocity analysis has shown that $\omega_3 = 2.094$ rad/s counter clockwise, $\omega_4 = 8.283$ rad/s counter clockwise, $\omega_5 = 2.188$ rad/s clockwise, and $\omega_6 = 7.770$ rad/s counter clockwise. Find the angular acceleration vectors for links 3, 4, 5, and 6. Also, find the current linear acceleration vectors for points P_1 and P_2.

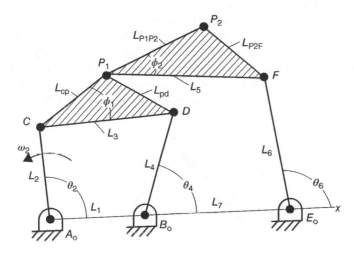

Figure 6.18   Problem 6.15

6.16   Given (Figure 6.19): $L_1 = 3.00$ in., $L_2 = 1.55$ in., $L_3 = 2.50$ in., $L_4 = 2.75$ in., $L_5 = 2.88$ in., $L_6 = 3.88$ in., $L_7 = 3.25$ in., $\theta_1 = 180°$, $\theta_7 = -44°$, and currently $\theta_2 = 103°$ and travel-ing at 8 rad/s counter clockwise. Link 2 is increasing speed at 2 rad/s^2. Also, $L_{cp} = 2.88$ in. and $L_{pd} = 4.25$ in. A positional analysis has determined that the angular positions of links 3, 4, 5, and 6 are $\theta_3 = 151.3°$, $\theta_4 = 80.4°$, $\theta_5 = 130.9°$, and $\theta_6 = 72.6°$. Similarly, the current location of point P is (4.61, 3.62) in. Also, a velocity analysis has shown that $\omega_3 = 2.018$ rad/s clockwise, $\omega_4 = 3.562$ rad/s counter clockwise, $\omega_5 = 3.579$ rad/s counter clockwise, and $\omega_6 = 0.00900$ rad/s counter clockwise. Find the angular acceleration vec-tors for links 3, 4, 5, and 6. Also, find the current linear acceleration vector for point P.

6.17   Given (Figure 6.19): $L_1 = 75.0$ mm, $L_2 = 40.0$ mm, $L_3 = 60.0$ mm, $L_4 = 70.0$ mm, $L_5 = 72.0$ mm, $L_6 = 95.0$ mm, $L_7 = 80.0$ mm, $\theta_1 = 180°$, $\theta_7 = -43°$, and currently $\theta_2 = 102°$ and cur-rently traveling at 12 rad/s counter clockwise. Link 2 is decreasing speed at 2 rad/s^2. Also, $L_{cp} = 72.0$ mm and $L_{pd} = 105$ mm. A positional analysis has determined that the angular positions of links 3, 4, 5, and 6 are $\theta_3 = 150.7°$, $\theta_4 = 78.2°$, $\theta_5 = 130.2°$, and $\theta_6 = 71.6°$. Sim-ilarly, the current location of point P is (117, 90.6) mm. Also, a velocity analysis has shown that $\omega_3 = 3.391$ rad/s clockwise, $\omega_4 = 5.399$ rad/s counter clockwise, $\omega_5 = 5.693$ rad/s counter clockwise, and $\omega_6 = 0.2023$ rad/s clockwise. Find the angular acceleration vectors for links 3, 4, 5, and 6. Also, find the current linear acceleration vector for point P.

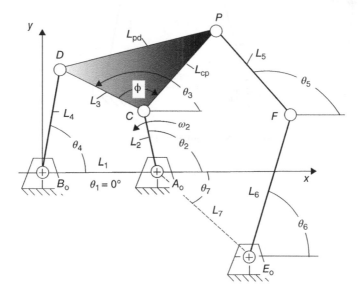

Figure 6.19   Problems 6.16 and 6.17

Programming Exercises

P6.1   Given (Figure 6.20): $L_1 = 7.25$ in., $L_2 = 2.00$ in., $L_3 = 7.00$ in., $L_4 = 4.50$ in., and $\theta_1 = 5°$.
       Start with $\theta_2 = 5°$ and increment it by $10°$ for a complete revolution. Link 2 is traveling at
       a constant 10 rad/s counterclockwise. Determine the angular acceleration vectors for
       links 3 and 4 for each θ_2 value.

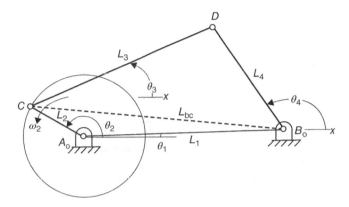

Figure 6.20   Programming exercise P6.1

P6.2   Given: $L_1 = 3.52$ in., $L_2 = 1.25$ in., $L_3 = 4.00$ in., $L_4 = 2.00$ in., $\theta_1 = 0°$, $L_{cp} = 3.00$ in., and
       $l_{pd} = 2.00$ in. Assume link 2 is traveling at a constant 12 rad/s counterclockwise. Start
       with $\theta_2 = 0°$ and increment it by $5°$ for a complete revolution. Find the angular acceler-
       ation vectors for links 3 and 4 assuming Figure 6.21 is drawn roughly to scale. Also,-
       determine the linear acceleration vector of point P. Plot the magnitude of the
       acceleration of point P versus the input angle θ_2.

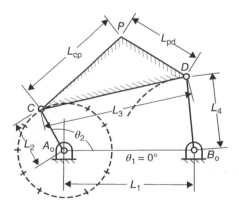

Figure 6.21   Programming exercise 2

P6.3   Given: $L_1 = 70.0$ mm, $L_2 = 55.0$ mm, $L_3 = 165$ mm, $\theta_1 = 270°$, and $\theta_4 = 0°$. Assume link 2 is traveling at a constant 20 rad/s counter clockwise. Start with $\theta_2 = -18.5°$ and increment it by 10° for a complete revolution. Find the angular acceleration vector for link 3 and the linear acceleration vector for link 4 (point D) assuming Figure 6.22 is drawn roughly to scale.

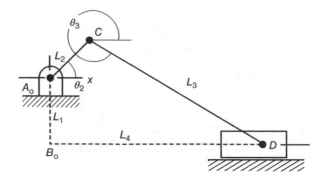

Figure 6.22   Programming exercise P6.3

P6.4   Given: $L_1 = 30.0$ mm, $L_2 = 55.0$ mm, $L_3 = 210$ mm, $\theta_1 = 270°$, $\theta_4 = 0°$, $L_{cp} = 84.0$ mm, and $L_{pd} = 170$ mm. Assume link 2 is traveling at a constant 10 rad/s clockwise. Start with $\theta_2 = 360°$ and decrement it by 5° for a complete revolution. Find the angular acceleration vector of link 3 and the linear acceleration vector of link 4 (point D) assuming Figure 6.23 is drawn roughly to scale. Also, find the acceleration vector (a_{px}, a_{py}) for point P for a complete revolution of link 2.

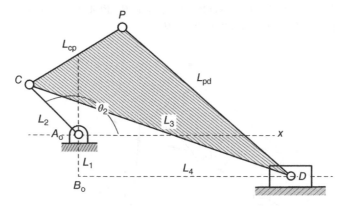

Figure 6.23   Programming exercise P6.4

P6.5   Given: $L_1 = 4.00$ in., $L_2 = 2.50$ in., $L_5 = 3.75$ in., $\theta_1 = \theta_5 = 0°$, and $\theta_3 = \theta_4$. Link 2 is traveling at a constant 16 rad/s counter clockwise. Start with $\theta_2 = 0°$ and increment it by 10°
for a complete revolution. Find the angular acceleration vector for link 4, the linear
acceleration vector of L_3 along link 4 (B_o to point C), and the vertical acceleration vector
of point D assuming Figure 6.24 is drawn roughly to scale. Plot the vertical acceleration
vector of point D versus the input angle θ_2.

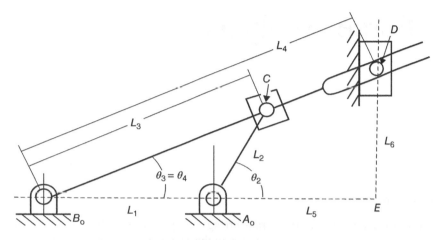

Figure 6.24   Programming exercise P6.5

P6.6   Given: $L_1 = 3.00$ in., $L_2 = 1.38$ in., $L_3 = 2.50$ in., $L_4 = 2.75$ in., $L_5 = 4.75$ in., $L_6 = 4.88$ in.,
$L_7 = 3.25$ in., $\theta_1 = 180°$, $\theta_7 = -44°$, $L_{cp} = 2.88$ in., and $L_{pd} = 4.25$ in. Link 2 is traveling at
a constant 10 rad/s counter clockwise. Start with $\theta_2 = 0°$ and increment it by 12° for a
complete revolution. Find the angular acceleration vectors for links 3, 4, 5, and 6
assuming Figure 6.25 is drawn roughly to scale. Also, find the acceleration vector
(a_{px}, a_{py}) for point P for a complete revolution of link 2.

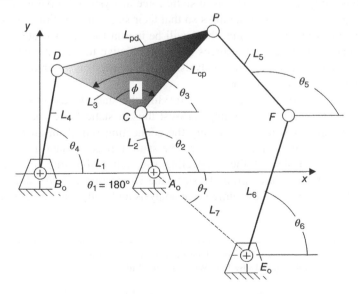

Figure 6.25   Programming exercise P6.6

7

Static Force Analysis

7.1 Introduction

A machine is a device that performs work by transferring energy by means of mechanical forces from a power source to a driven load. It is necessary in the design of a mechanism to know the manner in which forces are transmitted from the input to the output so that the components of the mechanism can be properly sized to withstand the stresses induced. If the links are not designed to be strong enough, failure will occur during operation. If the machine is overdesigned to have much more strength than required, then it may not be competitive with other machines in terms of cost, size, weight, or power requirements.

Various assumptions must be made during the force analysis that will reflect the specific characteristics of the mechanism being investigated. These assumptions should be verified as the design proceeds. For the initial analysis, a static force analysis will give us some idea of the forces that are present on each of the links so that their size can be estimated. All links have mass, and if the links are accelerating, there will be inertia forces associated with this motion. If the magnitudes of these inertia forces are small relative to the externally applied loads, then they can be neglected in the force analysis. Such an analysis is referred to as a static force analysis and is the topic of this chapter.

For example, during the normal operation of a forklift, the static loads on the forks far exceed any dynamic loads due to accelerating masses; thus, a static force analysis can be justified. On the other hand, if you are analyzing the connecting rod of an air compressor running at high speed, the inertia forces far exceed the static forces. A static force analysis could be used for the initial size of the connecting rod, but a dynamic force analysis would be required to verify you have properly sized the connecting rod. An analysis that includes the inertia effects is called a dynamic force analysis and will be covered in the next chapter.

Design and Analysis of Mechanisms: A Planar Approach, First Edition. Michael J. Rider.
© 2015 John Wiley & Sons, Ltd. Published 2015 by John Wiley & Sons, Ltd.

Another assumption that is often made is that frictional forces are negligible. Friction is inherent in all devices, and its degree is dependent upon many factors, including types of bearings, lubrication, loads, environmental conditions, etc. Friction will be neglected initially while sizing the parts but should be included in the final analysis if values are known.

Another assumption deals with the rigidity of the linkage components. No material is truly rigid, and all material will experience significant deformation if the forces are great enough. It is assumed that the mechanisms in this chapter are rigid and do not deform. The subject of mechanical vibrations due to deformation of a component is beyond the scope of this book.

7.2 Forces, Moments, and Free Body Diagrams

A force is a vector quantity; it has a magnitude with units and a direction. In 2-D, a force can be represented by $\overline{F} = (F_x, F_y)$, where F_x and F_y are the components of the force in the x and y directions.

A torque or moment, T, is defined as the moment of the force about a point and is also a vector quantity. We can calculate the torque or moment using the cross product, where $\overline{T} = \overline{r} \times \overline{F}$ or $\overline{T} = r \cdot F \cdot \sin\theta_{\text{between}}$ ($\perp$ to $\overline{r}$ and $\overline{F}$).

Free body diagrams (FBD) are extremely important and useful in a force analysis (Figure 7.1). A FBD is a sketch of a part showing all forces and moments or torques acting on it. Generally, the first step in a successful force analysis is the identification of the FBD for each of the parts used in the mechanism with the appropriate force, moments, angles, and distances defined (see Figure 7.2). Link 3 is a two-force member.

For a part to be in static equilibrium, the vector sum of all forces acting on the part must be zero and the vector sum of all moments about an arbitrary point must also be zero. There are many situations where the loading is essentially planar; thus, forces can be described by 2-D vectors. If the XY plane is designated as the plane of loading, then the static force analysis equations are

$$\rightarrow \sum F_x = 0 \tag{7.1}$$

$$\uparrow \sum F_y = 0 \tag{7.2}$$

$$(\text{c.c.w.}) \sum M_z = 0 \tag{7.3}$$

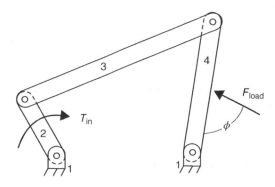

Figure 7.1   Four-bar linkage

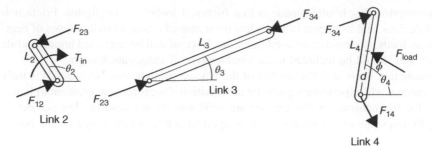

Figure 7.2   FBD of each moving link

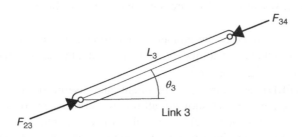

Figure 7.3   Two-force member

Since link 3 is a two-force member (forces applied at just two locations), we know from a course in statics that the force at each application point must be pointed toward the other force application point (see Figure 7.3). The only way the sum of the moments about any point can be zero is if the two forces are along the same line and in opposite directions. The same goes for the summation of forces in any direction. Therefore, F_{23} is at angle θ_3, and F_{34} (same magnitude as F_{23}) is at angle $(\theta_3 + 180°)$. If link 3 was not a two-force member, then this statement would not be true.

When we look at the FBD for link 2 (Figure 7.4), we see that F_{12} and F_{23} must be the same magnitude and in opposite directions; otherwise, the vector sum of the forces would not equal zero. If you didn't realize this, then force F_{12} should be written as its x and y components, F_{12x} and F_{12y}. Newton's third law states that for each action, there is an equal and opposite reaction. Therefore, F_{23} on link 2's FBD must be at an angle of $(\theta_3 + 180°)$. This makes F_{12} at an angle of θ_3. Forces F_{12} and F_{23} form a couple moment whose magnitude is equal to one of the forces times the perpendicular distance between the two forces. Note that the third term in the moment equation contains a negative sign because the $\cos(\theta_2)$ is negative and the force component creates a positive moment. Summing moments about the fixed bearing leads to

$$(\text{c.c.w.})\ \sum M_{12} = 0$$

$$-T_{in} + F_{23}\cos(\theta_3)L_2\sin\theta_2 - F_{23}\sin(\theta_3)L_2\cos\theta_2 = 0$$

$$-T_{in} + F_{23}L_2\sin(\theta_2 - \theta_3) = 0$$

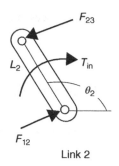

Link 2

Figure 7.4   FBD of link 2

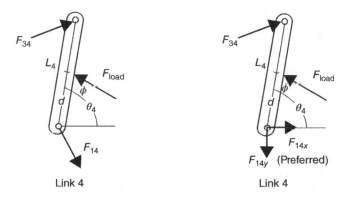

Link 4   Link 4

Figure 7.5   FBD of link 4

It may appear that the sign on the third term should be plus because the vertical component of the F_{23} force generates a counterclockwise (c.c.w.) moment, but you have to realize that the cosine of θ_2 in the second quadrant is negative so the third term needs a negative sign to produce a positive moment.

Another way of looking at the moment created by force, F_{23}, would be to use the cross product definition, $\bar{T} = \bar{r}x\bar{F}$. Note that $\bar{r} = (L_2 \cos\theta_2, L_2 \sin\theta_2)$ and $\bar{F} = (-F_{23}\cos\theta_3, -F_{23}\sin\theta_3)$. Taking the cross product leads to the same result shown previously:

$$\bar{r}x\bar{F} = \begin{vmatrix} i & j & k \\ L_2\cos\theta_2 & L_2\sin\theta_2 & 0 \\ -F_{23}\cos\theta_3 & -F_{23}\sin\theta_3 & 0 \end{vmatrix} = L_2 F_{23}(-\cos\theta_2 \sin\theta_3 + \cos\theta_3 \sin\theta_2)$$

$$\bar{r}x\bar{F} = +L_2 F_{23}\sin(\theta_2 - \theta_3)$$

Finally, let us examine the FBD of link 4 (see Figure 7.5). We know the direction of force F_{34}. We are typically given the direction of the load force, F_{load}. What we don't know is the direction of the force, F_{14}. In this case, it is better to show the x and y components of force, F_{14}.

This will allow the force F_{14x} to be in the sum of the forces in the x-direction and F_{14y} to be in the sum of the forces in the y-direction equation:

$$\rightarrow \sum F_x = 0$$
$$F_{34}\cos\theta_3 + F_{14x} + F_{load}\cos(\theta_4 + \phi) = 0$$
$$\uparrow \sum F_y = 0$$
$$F_{34}\sin\theta_3 - F_{14y} + F_{load}\sin(\theta_4 + \phi) = 0$$
$$(\text{c.c.w.}) \quad \sum M_{14} = 0$$
$$F_{load}d\sin\phi - F_{34}L_4\sin(\theta_4 - \theta_3) = 0$$

Example 7.1   Static force analysis of a 4-bar linkage

Problem: For the 4-bar linkage shown in Figure 7.6, determine the forces at each of the pins and the input torque for static equilibrium. The two fixed bearings are 10 in. apart on a horizontal surface. The input link, link 2, is 5 in. long and currently at 115°. The coupler link, link 3, is 15 in. long. The output link, link 4, is 10 in. long. The external force, F_{load}, is 40 lbs, is located 5 in. from the fixed bearing, and oriented 70° relative to link 4 as shown.

Solution:

1. Draw the FBD for each link (see Figures 7.2 through 7.5).
2. Write the vector loop position equation.
3. Write the static equilibrium equations for each link as needed (see Figure 7.7).
4. Guess initial values for θ_3 and θ_4. From sketch, let $\theta_3 = 20°$; $\theta_4 = 80°$.
5. Solve the set of equations (see Figure 7.8).
6. Check your work and verify that the answers seem reasonable.
7. Box in your answers with appropriate units and directions as appropriate.

Answers: $\boxed{F_{12} = 22.1 \ \text{lbs} = F_{23} = F_{34}}$, $\boxed{F_{14} = 31.5 \ \text{lbs}}$, and $\boxed{T_{in} = 110 \ \text{in. lbs c.w.}}$ Seems reasonable.

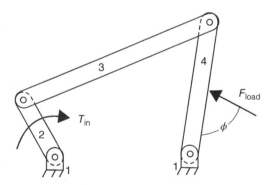

Figure 7.6   Four-bar linkage

Equations entered into EES

"Given data"
L_1=10
L_2=5
L_3=15
L_4=10
d=5
F_load=40
theta_2=115
phi=70

"Position analysis"
L_2*cos(theta_2)+L_3*cos(theta_3)−L_4*cos(theta_4)−L_1=0
L_2*sin(theta_2)+L_3*sin(theta_3)−L_4*sin(theta_4)=0

"Static force analysis"
F_load*d*sin(phi)−F_34*L_4*sin(theta_4−theta_3)=0
F_load*cos(theta_4+phi)+F_14x+F_34*cos(theta_3)=0
F_load*sin(theta_4+phi)+F_14y+F_34*sin(theta_3)=0
F_23=F_34
F_12=F_23
−T_in+F_23*L_2*sin(theta_2−theta_3)=0
F_14=sqrt(F_14^x2+F_14y^2)

Figure 7.7   Equations entered into EES

Solution
Unit settings: Eng F psia mass deg
d = 5 [in] F_{12} = 22.09 [lb.]
F_{14} = 31.52 [lb.] F_{14x} = 13.57 [lb.]
F_{14y} = 28.44 [lb.] F_{23} = 22.09 [lb.]
F_{34} = 22.09 [lb.] F_{load} = 40 [lb.]
L_1 = 10 [in] L_2 = 5 [in]
L_3 = 15 [in] L_4 = 10 [in]
ϕ = 70 [deg] θ^2 = 115 [deg]
θ^3 = 20.61 [deg] θ^4 = 78.89 [deg]
T_{in} = 110.1 [in.lb.]

Figure 7.8   Static force solution from EES

This is the solution at just one orientation of the input link, link 2. In order to obtain the largest values at each pin and the largest required input torque, this analysis needs to be repeated for each orientation of the input link. A MATLAB program or a table in EES can be used.

7.3   Multiforce Members

If a link has more than two forces associated with it, then it is better to define the forces at the connection points in terms of their x and y components (see Figure 7.9). The upper arm of the forklift has forces applied in four different locations. The applied load, P, is directed downward at the scoop. However, the directions for the three pin connections are unknown. In this case, the forces are shown as its horizontal and vertical components at each pin. The direction

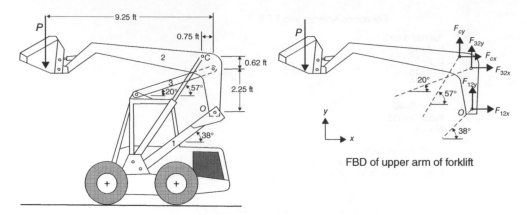

Figure 7.9 Multiforce members

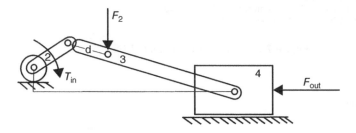

Figure 7.10 Slider–crank mechanism

assumed is not important. If you assume the wrong direction, the magnitude of the force component will come out negative.

The following slider–crank mechanism has two external forces applied: one at the slider and one on the connecting link, link 3 (see Figure 7.10). Link 3 is no longer a two-force member so we do not know the directions of the pin forces at either of its ends. Since we do not know the direction of the pin force between links 2 and 3, we do not know the direction of the pin force at the fixed bearing on the left. The direction of the horizontal and vertical force components is arbitrary, but once a component is drawn, the next time it shows up on a FBD it must be drawn in the opposite direction.

The FBD for link 2 (Figure 7.11) and its static equilibrium equations follows:

$$\rightarrow \sum F_x = 0$$

$$-F_{12x} + F_{23x} = 0$$

$$\uparrow \sum F_y = 0$$

$$-F_{12y} + F_{23y} = 0$$

$$(\text{c.c.w.}) \ \sum M_{12} = 0$$

$$-T_{in} + F_{23y}L_2 \cos\theta_2 - F_{23x}L_2 \sin\theta_2 = 0$$

Next, we draw the FBD for link 3 (see Figure 7.12).

However, if we draw link 3 so that it is located in the first quadrant where the sine and cosine are both positive, it is easier to write the static equilibrium equations (see Figure 7.13). Note that F_{23x} and F_{23y} are in the opposite direction compared to these forces shown on link 2's FBD:

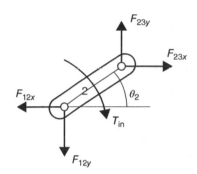

Figure 7.11   FBD of link 2

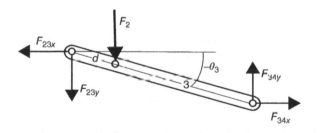

Figure 7.12   Original FBD of link 3

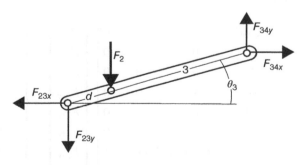

Figure 7.13   FBD of link 3 in 1st quadrant

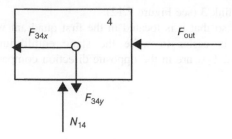

Figure 7.14 FBD of slider, link 4

$$\rightarrow \sum F_x = 0$$
$$-F_{23x} + F_{34x} = 0$$
$$\uparrow \sum F_y = 0$$
$$-F_{23y} + F_{34y} - F_2 = 0$$
$$(\text{c.c.w.}) \quad \sum M_{23} = 0$$
$$-F_2 d\cos\theta_3 + F_{34y} L_3 \cos\theta_3 - F_{34x} L_3 \sin\theta_3 = 0$$

Now, draw the FBD for the slider, link 4 (Figure 7.14), and then write the static equilibrium equations:

$$\rightarrow \sum F_x = 0$$
$$-F_{34x} - F_{\text{out}} = 0$$
$$\uparrow \sum F_y = 0$$
$$-F_{34y} + N_{14} = 0$$

We now have eight linear equations and eight unknowns. Summing moments about the pin would locate the normal force, N_{14}, which is not needed. None of the unknowns are raised to a power or in a trigonometric function; thus, the system of equations is linear. MATLAB is an easy way to solve simultaneous linear equations. The first step is to rewrite the eight equations in matrix form as shown below:

$$
\begin{bmatrix}
-1 & 0 & 0 & -L_2\sin\theta_2 & L_2\cos\theta_2 & 0 & 0 & 0 \\
0 & -1 & 0 & 1 & 0 & 0 & 0 & 0 \\
0 & 0 & -1 & 0 & 1 & 0 & 0 & 0 \\
0 & 0 & 0 & -1 & 0 & 1 & 0 & 0 \\
0 & 0 & 0 & 0 & -1 & 0 & 1 & 0 \\
0 & 0 & 0 & 0 & 0 & -1 & 0 & 0 \\
0 & 0 & 0 & 0 & 0 & -L_3\sin\theta_3 & L_3\cos\theta_3 & 1 \\
0 & 0 & 0 & 0 & 0 & 0 & -1 & 1
\end{bmatrix}
\begin{Bmatrix}
T_{\text{in}} \\
F_{12x} \\
F_{12y} \\
F_{23x} \\
F_{23y} \\
F_{34x} \\
F_{34y} \\
N_{14}
\end{Bmatrix}
=
\begin{Bmatrix}
0 \\
0 \\
0 \\
0 \\
F_2 \\
F_{\text{out}} \\
F_2 d\cos\theta_3 \\
0
\end{Bmatrix}
$$

Example 7.2   Static force analysis of a slider–crank mechanism

Problem: For the slider–crank mechanism shown in Figure 7.15, determine the forces at each of the pins and the input torque for static equilibrium. The fixed bearing is 30 mm above the slider pin. The input link, link 2, is 100 mm long and currently at 35°. The coupler link, link 3, is 350 mm long. The external force, F_{out}, is 100 N. The vertical force, F_2, is 64 N and is located 40 mm from the bearing between links 2 and 3.

Solution:

1. Draw the FBD for each link (see Figures 7.11 through 7.14).
2. Write the closed-loop position equations for a slider–crank mechanism.
3. Write the static equilibrium equations for each link as needed (see Figure 7.16).
4. Solve the set of equations (see Figure 7.17).
5. Check your work and verify that the answers seem reasonable.
6. Box in your answers with appropriate units and directions as appropriate.

(Negative answer indicates wrong direction assumed.)

$$F_{12x} = -100 \text{ N.}; \quad F_{12y} = -30.91 \text{ N.}$$

$$F_{12} = \sqrt{F_{12x}{}^2 + F_{12y}{}^2} = \sqrt{(-100)^2 + (-30.91)^2} = 104.7 \text{ N.}$$

$$F_{23x} = -100 \text{ N.}; \quad F_{23y} = -30.91 \text{ N.}$$

$$F_{23} = \sqrt{F_{23x}{}^2 + F_{23y}{}^2} = \sqrt{(-100)^2 + (-30.91)^2} = 104.7 \text{ N.}$$

$$F_{34x} = -100 \text{ N.}; \quad F_{34y} = 33.09 \text{ N.}$$

$$F_{34} = \sqrt{F_{34x}{}^2 + F_{34y}{}^2} = \sqrt{(-100)^2 + 33.09^2} = 105.3 \text{ N.}$$

Answers: $\boxed{\theta_3 = -14.4°}$, $\boxed{T_{in} = 3.20 \text{ N m c.c.w.}}$, $\boxed{F_{12} = 105 \text{ N} = F_{23}}$, and $\boxed{F_{34} = 105 \text{ N}}$.

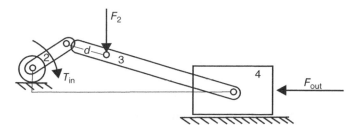

Figure 7.15   Slider–crank mechanism

```
fomat compact; format bank
L1=.030;
L2=.100;
L3=.350;
theta2=35;
theta3=asin((-L1-L2*sind(theta2))/L3)*180/pi
L4=L2*cosd(theta2)+L3*cosd(theta3);
Fout=100;
F2=64;
d=.040;

A=[-1 0 0 -L2*sind(theta2) L2*cosd(theta2); 0 0 0;...
    0 -1 0 1 0 0 0 0;...
    0 0 -1 0 1 0 0 0;...
    0 0 0 -1 0 1 0 0;...
    0 0 0 0 -1 0 1 0;...
    0 0 0 0 0 -1 0 0;...
    0 0 0 0 0 -L3*sind(theta3) L3*cosd(theta3) 0;...
    0 0 0 0 0 0 -1 1];
RHS=[0; 0; 0; 0; F2; Fount; F2*d*cosd(theta3); 0];

Forces=A\RHS
```

Figure 7.16   MATLAB code

```
>> Eight_Eqns
theta3 =
         -14.45
Forces =
            3.20
         -100.00
          -30.91
         -100.00
          -30.91
         -100.00
           33.09
           33.09
```

Figure 7.17   MATLAB answers

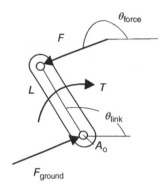

Figure 7.18 General link and force orientation

It would appear that the input torque needs to be clockwise (c.w.) to counteract the force applied on the slider; however, because of force, F_2, the actual input torque in this configuration needs to be counterclockwise.

7.4 Moment Calculations Simplified

Sign errors can occur in the moment equation when taking the moment about a point when the link is not in the first quadrant because the sine or cosine in some of the other quadrants is negative. The following procedure works 100% of the time and will always provide the correct sign for each moment term.

As previously noted in this book, the vector angular orientation is always defined from the $+x$-axis with counterclockwise being positive. With this in mind, a general equation can be derived for any $\bar{r}x\bar{F}$ moment term. To calculate the moment about point A_o caused by force, F, do the following (see Figure 7.18):

$$\bar{L}=L \angle \theta_{link} = (L\cos\theta_{link}, L\sin\theta_{link})$$

$$\bar{F}=F \angle \theta_{force} = (F\cos\theta_{force}, F\sin\theta_{force})$$

$$\bar{r}x\bar{F}= \begin{vmatrix} i & j & k \\ L\cos\theta_{link} & L\sin\theta_{link} & 0 \\ F\cos\theta_{force} & F\sin\theta_{force} & 0 \end{vmatrix} = FL(\sin\theta_{force}\cos\theta_{link} - \cos\theta_{force}\sin\theta_{link})$$

$$\boxed{\bar{r}x\bar{F} = + FL\sin(\theta_{force} - \theta_{link})} \tag{7.4}$$

Thus, the moment caused by any force vector with the link vector in any quadrant is equal to $FL\sin(\theta_{force} - \theta_{link})$ as long as the force and link vectors are defined relative to the $+x$-axis.

Problems

7.1 Determine the required input torque on link 2 and the magnitude of the pin forces for the slider–crank mechanism shown in Figure 7.19. Given $L_1 = 2.10$ in., $L_2 = 2.00$ in., $L_3 = 6.50$ in., and load force $F_{load} = 100$ lbs to the left, is the required torque clockwise or counter clockwise?

a. $\theta_2 = 43°$, $\theta_3 = -32.2°$
b. $\theta_2 = 133°$, $\theta_3 = -33.2°$
c. $\theta_2 = 223°$, $\theta_3 = -6.50°$
d. $\theta_2 = 343°$, $\theta_3 = -13.5°$

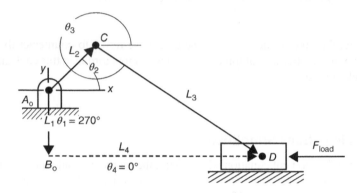

Figure 7.19 Problem 7.1

7.2   Determine the required input torque and the magnitude of the pin forces for the slider–
      crank mechanism shown in Figure 7.20. The link lengths of $L_1 = 0.00$ in., $L_2 = 2.50$ in.,
      and $L_3 = 4.30$ in. $= c = d$. The forces are constant at $F_1 = 100$ lbs $\leftarrow$, $F_2 = 50$ lbs $\leftarrow$, and
      $F_3 = 75$ lbs $\downarrow$. Is the required torque clockwise or counter clockwise?

   a. $\theta_2 = 30°$, $\theta_3 = -16.9°$
   b. $\theta_2 = 75°$, $\theta_3 = -34.2°$
   c. $\theta_2 = 115°$, $\theta_3 = -31.8°$
   d. $\theta_2 = 330°$, $\theta_3 = +16.9°$

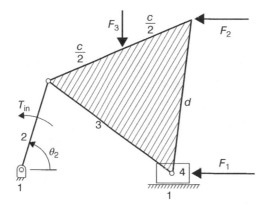

Figure 7.20   Problem 7.2

7.3 Determine the required input torque and the magnitude of the pin forces for the 4-bar linkage shown in Figure 7.21. The link lengths of $L_1 = 140$ mm, $L_2 = 40$ mm, $L_3 = 160$ mm, $L_4 = 120$ mm, $d_3 = 110$ mm, and $d_4 = 95$ mm. The forces are constant at $F_3 = 200$ N at $\phi = 60°$ away from link 3, and $F_4 = 150$ N always perpendicular to link 4. Is the required torque clockwise or counter clockwise?

a. $\theta_2 = 40°$, $\theta_3 = 35.3°$, $\theta_4 = 79.8°$
b. $\theta_2 = 80°$, $\theta_3 = 30.2°$, $\theta_4 = 87.5°$
c. $\theta_2 = 140°$, $\theta_3 = 33.5°$, $\theta_4 = 108.1°$
d. $\theta_2 = 225°$, $\theta_3 = 52.0°$, $\theta_4 = 125.5°$

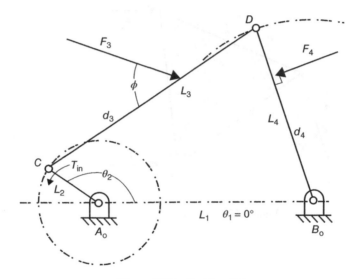

Figure 7.21   Problem 7.3

7.4   Determine the required input torque and the magnitude of the pin forces for the inverted slider–crank mechanism shown in Figure 7.22. Given $L_1 = 140$ mm, $L_2 = 85$ mm, $L_4 = 250$ mm, and load force $F_4 = 300$ N at a constant angle of $\phi = 48°$ from link 4 as shown, is the required torque clockwise or counter clockwise?

a. $\theta_2 = 66°$, $\theta_3 = 24.0°$, $L_3 = 191$ mm
b. $\theta_2 = 96°$, $\theta_3 = 32.8°$, $L_3 = 156$ mm
c. $\theta_2 = 136°$, $\theta_3 = 36.8°$, $L_3 = 98.5$ mm
d. $\theta_2 = 236°$, $\theta_3 = -37.3°$, $L_3 = 116.3$ mm

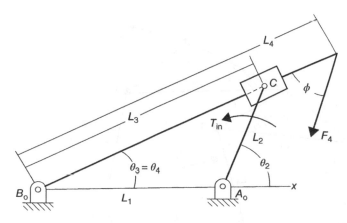

Figure 7.22   Problem 7.4

7.5   Determine the required input torque and the magnitude of the pin forces for the 6-bar linkage shown in Figure 7.23. Given: $L_1 = 2.88$ in., $L_2 = 2.75$ in., $L_3 = 4.00$ in., $L_4 = 3.25$ in., $L_5 = 4.75$ in., $L_6 = 4.00$ in., $L_7 = 4.38$ in., $\theta_1 = \theta_7 = 0°$, and currently $\theta_2 = 101°$. Also, $L_{cp} = 2.50$ in., $L_{pd} = 2.37$ in., $L_{P1P2} = 3.25$ in., and $L_{P2F} = 2.38$ in. Label the angles θ_3 and θ_5 on your sketch. A position analysis reveals that $\theta_3 = 7.2°$, $\theta_4 = 80.0°$, $\theta_5 = -6.2°$, $\theta_6 = 107.1°$, $P_{1x} = 1.364$ in., $P_{1y} = 4.337$ in., $P_{2x} = 4.400$ in., and $P_{2y} = 5.500$ in. The force F_1 is 100 lbs, is perpendicular to link 4, and is located 2.80 in. from fixed bearing B_o. The force F_2 is 160 lbs and is always located at $\phi_3 = 50°$ away from L_{P2F} as shown. The force F_3 is 80 lbs, is perpendicular to link 6, and is located 3.00 in. from fixed bearing E_o.

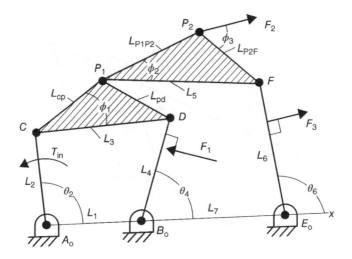

Figure 7.23   Problem 7.5

7.6   Determine the crushing force provided by the link 6 when an input torque of 200 N m is applied to link 2 as shown in Figure 7.24. What are the magnitudes of the pin forces when θ_2 is 194°? Given: $L_2 = 0.100$ m, $L_3 = L_4 = L_5 = 0.300$ m. Fixed bearing O_2 is 0.300 m from the origin. Fixed bearing O_4 is 0.400 m below the origin. Also, find θ_3, θ_4, and θ_5.

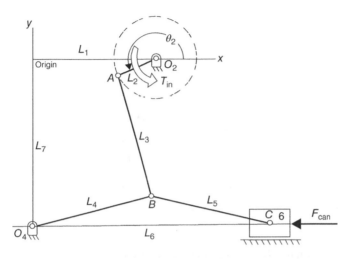

Figure 7.24   Problem 7.6

Programming Exercises

P7.1   Determine the required input torque on link 2 and the magnitude of the pin forces for the slider–crank mechanism shown in Figure 7.25 starting at $\theta_2 = 0°$ and incrementing it by 10° for a complete revolution. Given $L_1 = 2.10$ in., $L_2 = 2.00$ in., $L_3 = 6.50$ in., and load force $F_{load} = 100$ lbs to the left, is the required torque at each θ_2 clockwise or counter clockwise? Show the results in a table with θ_2 in the first column.

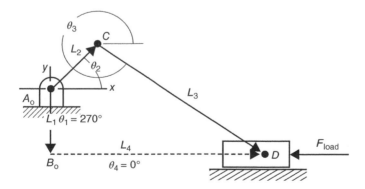

Figure 7.25   Programming exercise P7.1

P7.2   Determine the required input torque on link 2 and the magnitude of the pin forces
for the inverted slider–crank mechanism shown in Figure 7.26 starting at $\theta_2 = 0°$ and
incrementing it by 10° for a complete revolution. Given $L_1 = 140$ mm, $L_2 = 85$ mm,
$L_4 = 250$ mm, and load force $F_4 = 300$ N at an angle of $\phi = 48°$ from link 4 as shown,
is the required torque at each θ_2 clockwise or counter clockwise? Show the results in
a table with θ_2 in the first column.

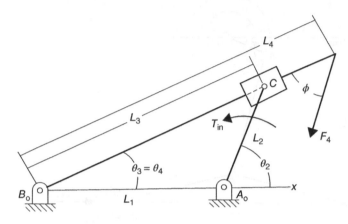

Figure 7.26   Programming exercise P7.2

P7.3 Determine the required input torque on link 2 and the magnitude of the pin forces for the 4-bar linkage shown in Figure 7.27 starting at $\theta_2 = 0°$ and incrementing it by 15° for a complete revolution. The link lengths of $L_1 = 140$ mm, $L_2 = 40$ mm, $L_3 = 160$ mm, $L_4 = 120$ mm, $d_3 = 110$ mm, and $d_4 = 95$ mm. The forces are constant at $F_3 = 200$ N at $\phi = 60°$ away from link 3, and $F_4 = 150$ N always perpendicular to link 4. Is the required torque at each θ_2 clockwise or counter clockwise? Show the results in a table with θ_2 in the first column.

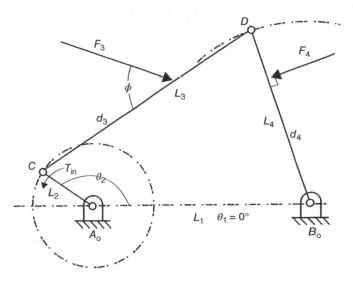

Figure 7.27   Programming exercise P7.3

P7.4 Determine the created can-crushing force and the magnitude of the pin forces for the 6-bar linkage shown in Figure 7.28 starting at $\theta_2 = 0°$ and incrementing it by 12° for a complete revolution. Given: $L_2 = 0.100$ m, $L_3 = L_4 = L_5 = 0.300$ m. Fixed bearing O_2 is 0.300 m from the origin. Fixed bearing O_4 is 0.400 m below the origin. A constant input torque of 200 N m is applied to link 2.

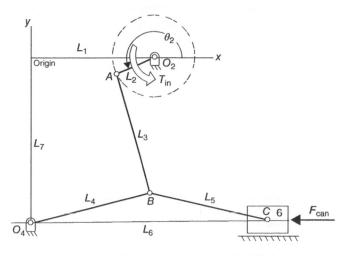

Figure 7.28   Programming exercise P7.4

8

Dynamic Force Analysis

8.1 Introduction

This chapter deals with dynamic forces in mechanisms. Dynamic forces are created by accelerating masses, and since most mechanisms contain accelerating masses, dynamic forces are present.

Let's assume we have a link rotating about a fixed pivot with a constant angular velocity. We know that in order to maintain a circular path of motion of the mass center, the link will experience a force acting radially outward. This force is commonly referred to as centrifugal force and is a dynamic force resulting from the centripetal acceleration of the mass center. This type of a dynamic force is very prevalent in mechanisms. For example, a rotating shaft will experience centrifugal force if its mass center does not lie exactly on the bearing's centerline. Differences between the center of mass and the center of rotation can occur because of manufacturing inaccuracies, inconsistent material density, and bowing or bending of the shaft.

The basis for investigating dynamic forces in mechanisms is Newton's second law of motion, which states that a particle acted on by a force whose resultant is not zero will move in such a way that the time rate of change of its momentum will at any instant be proportional to the resulting force. This law is expressed mathematically, for the special case of constant mass, as shown in equations 8.1 and 8.2, 8.3 or 8.4.

$$\sum \bar{F} = m\frac{d\bar{v}_G}{dt} = m\bar{a}_G \qquad (8.1)$$

and

$$\sum \overline{M_G} = I_G\frac{d\bar{\omega}}{dt} = I_G\bar{\alpha} \qquad (8.2)$$

or

$$\sum M_P = I_G\alpha - y_{G/P}ma_{GX} + x_{G/P}ma_{GY} \qquad (8.3)$$

Design and Analysis of Mechanisms: A Planar Approach, First Edition. Michael J. Rider.
© 2015 John Wiley & Sons, Ltd. Published 2015 by John Wiley & Sons, Ltd.

or

$$\sum M_P = I_P\alpha - y_{P/G}ma_{PX} + x_{P/G}ma_{PY} \tag{8.4}$$

where

G = center of gravity

P = any point on the object

Note that these are vector equations; thus, in 2-D, the sum of the forces can be done in two arbitrary perpendicular axes directions, and the moment is done about an axis perpendicular to the 2-D plane. There are three forms given for the sum of the moments. Any one of the three equations can be used depending upon the free body diagram (FBD). All three generate the same answer; thus, they are not independent of each other.

The mass moment of inertia of a particle about a particular axis is defined as the product of the mass of the particle times the square of the distance that the particle is away from the axis. For continuous bodies made up of an infinite number of mass particles, the mass moment of inertia, I, is the integral of all the individual mass moments of inertia of the particles:

$$I = \int_{mass} r^2 dm \tag{8.5}$$

This mathematical definition can be used to compute the mass moment of inertia of a mechanism's member. Tables are available in most mechanics texts that contain expressions for the moments of inertia for standard shapes. However, experimental methods of determination are often used for complex shapes, such as connecting rods or cams. In the SI, mass moment of inertia is usually expressed in units of kilogram meter squared (kg·m^2). In the English system, the units are slug foot squared (slug·ft^2) or inch-pound force second squared (in.·lbf·s^2).

The parallel axis theorem relates the mass moment of inertia about the center of gravity to the mass moment of inertia about any other point on the object:

$$I_{any\_axis} = I_G + md^2 \tag{8.6}$$

where

d = distance from center of gravity to any axis

In a dynamics course, Newton's equations are used to solve for the motion of a body resulting from the application of known forces and torques. In mechanism design, Newton's equations are used to solve for the forces and torques required to produce the known motion of the mechanism. The ideal motion of the mechanism is assumed, and the resulting forces and torques are calculated, and from this information, the size of the mechanism's members and its bearings and shafts are determined, along with the required input power. It is next to impossible to select design components so that the required forces for ideal motion are produced. Therefore, it may be necessary to evaluate the mechanism designed this way by either constructing and testing the physical prototype or by examining the computer model based on this approach and then fine-tuning the design.

8.2 Link Rotating about Fixed Pivot Dynamic Force Analysis

Figure 8.1 shows link 2 rotating in a vertical plane about fixed bearing O_2 at $\omega_2 = 10$ rad/s counter clockwise (c.c.w.) and $\alpha_2 = 72$ rad/s^2 counter clockwise Assume $I_o = 0.045$ slug·ft^2, $W_2 = 7$ lbs, $L_2 = 1$ ft, and $r_2 = 0.56$ ft. Determine the magnitude of the pin force at O_2 and the required input torque. Currently, θ_2 is 150° and nothing is attached at point A.

This is a typical dynamics problem which can be found in any engineering dynamics textbook, so let us follow the procedure learned in the dynamics course:

1. Draw the FBD of link 2 placing a horizontal and vertical force at O_2 and a weight at G_2.
2. Draw a kinetic diagram showing the angular velocity and acceleration of link 2 along with the normal and tangential acceleration vectors at the center of gravity.
3. Write Newton's laws of motion.
4. Solve the system of equations.
5. Verify that the answer makes sense.

Example 8.1   Perform a dynamic force analysis on a single rotating link

Problem: Figure 8.1 shows link 2 rotating in a vertical plane about fixed bearing O_2 at $\omega_2 = 10$ rad/s counter clockwise and $\alpha_2 = 72$ rad/s^2 counter clockwise. Assume $I_o = 0.045$ slug·ft^2, $W_2 = 7$ lbs, $L_2 = 1$ ft, and $r_2 = 0.56$ ft. Currently, θ_2 is 150° and nothing is attached at point A.

Find: Determine the magnitude of the pin force at O_2 and the required input torque.

Solution: From the FBD and the kinetic diagrams shown in Figure 8.2, write Newton's equations:

$$\rightarrow \sum F_x = ma_{Gx}$$
$$F_{O2X} = m_2\left(-a_{gn}\cos\theta_2 - a_{gt}\sin\theta_2\right)$$
$$\uparrow \sum F_y = ma_{Gy}$$
$$F_{O2Y} = m_2\left(-a_{gn}\sin\theta_2 + a_{gt}\cos\theta_2\right)$$
$$(+\text{c.c.w.}) \sum M_{O2} = I_{O2}\alpha - y_{O2/G}ma_{O2X} + x_{O2/G}ma_{O2Y}$$
$$T_{in} - W_2 r_2 \cos\theta_2 = I_{O2}\alpha_2$$

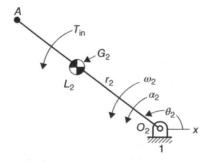

Figure 8.1   Rotating link about fixed pivot

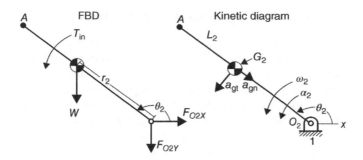

Figure 8.2 FBD and KD of link 2

It may appear that the sign on each of the cosine terms is incorrect, but one must remember that the cosine of an angle in the second quadrant is negative (cos 150° = −0.8660). Also, since point O_2 is a fixed bearing, the acceleration of O_2 is zero:

$$m_2 = \frac{W_2}{g} = \frac{7 \text{ lbs}}{32.2 \frac{\text{ft}}{s^2}} = 0.2174 \text{ slugs}$$

$$a_{gn} = \omega_2{}^2 r_2 = \left(10 \frac{\text{rad}}{s}\right)^2 (0.56 \text{ ft}) = 56 \frac{\text{ft}}{s^2}$$

$$a_{gt} = \alpha_2 r_2 = \left(72 \frac{\text{rad}}{s^2}\right)(0.56 \text{ ft}) = 40.32 \frac{\text{ft}}{s^2}$$

$$I_{O2} = 0.045 \text{ slug} \cdot \text{ft}^2$$

Substitute values into Newton's equations:

$$F_{O2X} = m_2 \left(-a_{gn} \cos\theta_2 - a_{gt} \sin\theta_2\right)$$

$$F_{O2X} = 0.2174 \text{ slugs} \left(-56 \frac{\text{ft}}{s^2} \cos 150° - 40.32 \frac{\text{ft}}{s^2} \sin 150°\right) = 6.160 \text{ lbs}$$

$$F_{O2Y} = m_2 \left(-a_{gn} \sin\theta_2 + a_{gt} \cos\theta_2\right)$$

$$F_{O2Y} = 0.2174 \text{ slugs} \left(-56 \frac{\text{ft}}{s^2} \sin 150° + 40.32 \frac{\text{ft}}{s^2} \cos 150°\right) = -13.68 \text{ lbs}$$

$$F_{O2} = \sqrt{F_{O2X}{}^2 + F_{O2Y}{}^2} = \sqrt{6.160^2 + (-13.68)^2} = 15.0 \text{ lbs}$$

$$T_{in} - W_2 r_2 \cos\theta_2 = I_{O2} \alpha_2$$

$$T_{in} = \left(0.045 \text{ slug} \cdot \text{ft}^2\right)\left(72 \frac{\text{rad}}{s^2}\right) + (7 \text{ lbs})(0.56 \text{ ft})\cos 150° = -0.155 \text{ ft} \cdot \text{lbs}$$

Note that a negative value indicates that the force or torque was assumed in the wrong direction on the FBD; thus, the input torque is clockwise (c.w.), and F_{OY2} is upward.

Answers: $\boxed{T_{in} = 0.155 \text{ ft} \cdot \text{lbs c.w.}}$ and $\boxed{F_{O2} = 15.0 \text{ lbs}}$.

8.3   Double-Slider Mechanism Dynamic Force Analysis

Next, we will examine a double-slider mechanism as shown in Figure 8.3.

Assume that slider 2 and slider 4 have a mass of 5 kg each. Connecting rod, link 3, has a mass of 3 kg and is 0.90 m long (L_3). The center of gravity for link 3 is 0.4 m ($L_{G3/2}$) from the pin connection of link 2. The radius of gyration about the center of gravity of link 3 is 0.370 m. Both sliders are on smooth surfaces; thus, friction can be neglected as this point. What is the required force, F_2, so that slider 4 accelerates at 5 m/s² to the left? What are the pin forces? Slider 4 is currently moving at 2 m/s to the left. Link 3 is currently at 130°. The FBD for link 3 is shown in the first quadrant (Figure 8.4), even though it was given in the second quadrant. Why?

There are seven unknowns present: N_{12}, F_{23X}, F_{23Y}, F_2, F_{34X}, F_{34Y}, and N_{14}. Since we are going to treat sliders 2 and 4 as particles, only Newton's sum of the force vector equation applies. Link 3 is treated as a rigid body so all three of Newton's 2-D equations apply. This gives us seven equations for the seven unknowns as long as we know all of the acceleration vectors for each link.

Using the relative acceleration equation from Chapter 6 allows us to determine the angular acceleration vector of link 3 and the linear acceleration vector for slider 2. A velocity analysis will provide us with the angular velocity of link 3 which is needed to calculate the normal

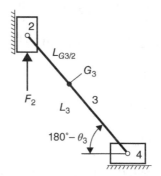

Figure 8.3   Double-slider mechanism

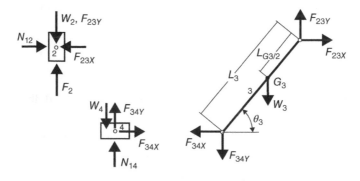

Figure 8.4   FBD of double slider

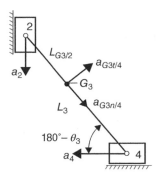

Figure 8.5   Kinetic diagram of double slider

acceleration vector at the center of gravity for link 3 (see Figure 8.5). A velocity and acceleration analysis leads to

$$\omega_3 = 2.901 \frac{\text{rad}}{s} \,(\text{c.w.})$$

$$\alpha_3 = 0.1910 \frac{\text{rad}}{s^2} \,(\text{c.w.})$$

$$a_{G2Y} = 5.691 \frac{\text{m}}{s^2} \downarrow = -a_2$$

For the center of gravity of link 3, we can use the relative acceleration equation

$$\overline{a_{G3}} = \overline{a_4} + \overline{a_{G3t/4}} + \overline{a_{G3n/4}}$$

$$\overline{a_4} = 5 \frac{\text{m}}{s^2} \angle 180°$$

$$\overline{a_{G3t/4}} = \alpha_3 (L_3 - L_{G3/2}) \angle (\theta_3 - 90°) = 0.1910(0.9 - 0.4) \angle 40°$$

$$\overline{a_{G3n/4}} = \omega_3^{\,2} (L_3 - L_{G3/2}) \angle (\theta_3 - 180°) = 2.901^2(0.9 - 0.4) \angle (-50°)$$

$$\overline{a_{G3}} = 3.865 \frac{\text{m}}{s^2} \angle -125.1°$$

$$\overline{a_{G3}} = (a_{G3x}, a_{G3y}) = (-2.222, -3.162) \frac{\text{m}}{s^2}$$

Now, we are ready to write the appropriate Newton's law equations for the three links. For link 4,

$$\rightarrow \sum F_x = m_4 a_{4x}$$

$$F_{34X} = 5 \text{ kg.} \left(-5 \frac{\text{m}}{s^2}\right) = -25 \text{ N}$$

$$\uparrow \sum F_y = m_4 a_{4y} = 0$$

$$F_{34Y} + N_{14} - W_4 = 0$$

For link 2,

$$\rightarrow \sum F_x = m_2 a_{2x} = 0$$

$$N_{12} - F_{23X} = 0$$

$$\uparrow \sum F_y = m_2 a_{2y}$$

$$F_2 - F_{23Y} - W_2 = 5 \text{ kg} \left(-5.691 \tfrac{\text{m}}{s^2}\right) = -28.46 \text{ N}$$

For link 3,

$$\rightarrow \sum F_x = m_3 a_{3x}$$

$$F_{23X} - F_{34X} = 3 \text{ kg} \left(-2.222 \tfrac{\text{m}}{s^2}\right) = -6.666 \text{ N}$$

$$\uparrow \sum F_y = m_3 a_{3y}$$

$$F_{23Y} - F_{34Y} - W_3 = 3 \text{ kg} \left(-3.162 \tfrac{\text{m}}{s^2}\right) = -9.486 \text{ N}$$

$$(\text{c.c.w.}) \sum M_{G3} = I_{G3} \alpha_3$$

$$\sum M_{G3} = F_{23Y} L_{G3/2} \cos\theta_3 - F_{23X} L_{G3/2} \sin\theta_3 - F_{34X} \left(L_3 - L_{G3/2}\right) \sin\theta_3 + F_{34Y} \left(L_3 - L_{G3/2}\right) \cos\theta_3$$

$$I_{G3} \alpha_3 = \left(k_{G3}^2 m_3\right) \alpha_3 = (0.370 \text{ m})^2 (3 \text{ kg}) \left(-0.1910 \tfrac{\text{rad}}{s^2}\right) = -0.7831 \text{ N} \cdot \text{m}$$

The pin forces are

$$F_{23} = \sqrt{F_{23X}^2 + F_{23Y}^2}$$

$$F_{34} = \sqrt{F_{34X}^2 + F_{34Y}^2}$$

Solving the system of linear equations using a calculator, MS Excel, EES, or MATLAB leads to

$$F_2 = 66.4 \text{ N}$$

$$F_{23} = 55.7 \text{ N}$$

$$F_{34} = 35.9 \text{ N}$$

The same procedure can be used for a slider–crank mechanism or a four-bar linkage. Simply draw the FBD for each moving link along with its kinetic diagram. Perform a position, velocity, and acceleration analysis on the device. Calculate the acceleration vectors for all mass centers. Write the appropriate Newton's law equations. Solve for the unknown forces and torques.

Problems

8.1   For the slider–crank mechanism shown in Figure 8.6, determine the required torque on link 2 necessary to keep link 2 moving at a constant speed of 1000 r.p.m. counter clockwise. Also, what are the magnitudes of the pin forces at pins A, B, and G_2? Link 2, currently at 30°, has a mass of 10 kg and its center of gravity is located at the fixed pivot. The distance from G_2 to A is 0.28 m. Link 3 has a mass of 6 kg and its center of gravity is located 0.25 m from pin A. Link 3 is 1.0 m long with a radius of gyration about the center of gravity of 0.456 m. The piston has a mass of 5 kg. The force applied to the piston is 500 N. What are the three pin forces at G_2, A, and B? Currently, $\theta_3 = 104°$, $\omega_3 = 15.11$ rad/s clockwise, $V_4 = V_B = 29.06$ m/s (↑), $\alpha_3 = 2684$ rad/s² clockwise, and $a_4 = a_B = 1106$ m/s² (↓ down).

8.2   For the slider–crank mechanism shown in Figure 8.6, determine the required torque on link 2 necessary to keep link 2 moving at a constant speed of 200 r.p.m. counter clockwise. Also, what are the magnitudes of the pin forces at pins A, B, and G_2? Link 2, currently at 30°, has a weight of 10 lbs and its center of gravity is located at the fixed pivot. The distance from G_2 to A is 0.28 ft. Link 3 has a weight of 6 lbs and its center of gravity is located 0.25 ft from pin A. Link 3 is 1.0 ft long with a radius of gyration about the center of gravity of 0.456 ft. The piston has a weight of 5 lbs. The force applied to the piston is 500 lbs. What are the three pin forces at G_2, A, and B? Currently, $\theta_3 = 104°$, $\omega_3 = 3.022$ rad/s clockwise, $V_4 = V_B = 5.812$ ft/s (↑), $\alpha_3 = 107.4$ rad/s² clockwise, and $a_4 = a_B = 44.24$ ft/s² (↓ down).

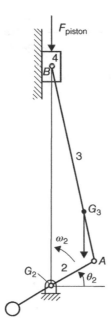

Figure 8.6   Problems 8.1 and 8.2

8.3 Determine the required input torque on link 2 and the magnitude of the pin forces for the slider–crank mechanism shown in Figure 8.7. Given $L_1 = 2.10$ in., $L_2 = 2.00$ in., $L_3 = 6.50$ in., and load force $F_{load} = 100$ lbs to the left, is the required torque clockwise or counter clockwise? Link 2 weighs 3.22 lbs and is currently rotating at a constant 30 rad/s clockwise. Link 3 weighs 6.44 lbs and link 4 weighs 4.83 lbs. Links 2 and 3 can be treated as slender rods with their mass center at their geometric center:

a. $\theta_2 = 43°$, $\theta_3 = -32.2°$, $\omega_3 = 7.978$ rad/s counter clockwise, $\alpha_3 = 183.1$ rad/s^2 counter clockwise

b. $\theta_2 = 133°$, $\theta_3 = -33.2°$, $\omega_3 = 7.527$ rad/s clockwise, $\alpha_3 = 205$ rad/s^2 counter clockwise

c. $\theta_2 = 223°$, $\theta_3 = -6.50°$, $\omega_3 = 6.795$ rad/s clockwise, $\alpha_3 = 195.3$ rad/s^2 clockwise

d. $\theta_2 = 343°$, $\theta_3 = -13.5°$, $\omega_3 = 9.078$ rad/s counter clockwise, $\alpha_3 = 103$ rad/s^2 clockwise

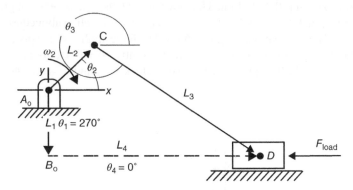

Figure 8.7   Problem 8.3

8.4   For the vertical standing 4-bar linkage shown in Figure 8.8, determine the required
      input torque on link 2 so that link 2 rotates at a constant speed of 100 r.p.m.
      clockwise. Also, what are the magnitudes of the pin forces at pins A, B, O_2, and O_4?
      Link 2, currently at 145°, is 0.40 m long with a mass of 0.60 kg. Link 3 is 1.20 m
      long with a mass of 1.80 kg. Link 4 is 0.80 m long with a mass of 1.2 kg. All three
      links can be treated as slender rods with their mass center at their geometric center
      as shown. The 4-bar is drawn roughly to scale. Fixed bearing O_4 is 1.2 m (L_1) right
      of fixed bearing O_2. Currently, $\theta_3 = 22.2°$, $\theta_4 = 121.4°$, $\omega_3 = 1.416$ rad/s clockwise,
      $\omega_4 = 4.459$ rad/s clockwise, $\alpha_3 = 20.18$ rad/s^2 counter clockwise, and $\alpha_4 = 23.82$ rad/s^2
      clockwise.

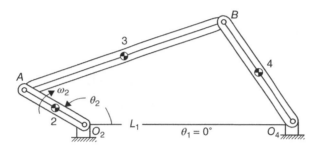

Figure 8.8   Problem 8.4

8.5   For the 4-bar linkage shown in Figure 8.9, determine the required input torque on link 2 and the magnitude of the pin forces at O_2, O_4, A, and B for the given conditions. Link 2 (O_2A) is 290 mm long and at an angle of $\theta_2 = 200°$. Link 2 has a mass of 3.1 kg and a mass moment of inertia of 0.026 kg·m² about its mass center, which is 150 mm from O_2. Link 2 is currently rotating at 3.00 rad/s clockwise and accelerating at 1.60 rad/s². Fixed bearing O_4 is located 209 mm at 133.5° from fixed bearing O_2. Link 4 (O_4B) is 390 mm long and at an angle of $\theta_4 = 220°$. Link 4 has a mass of 4.1 kg and a mass moment of inertia of 0.060 kg·m² about its mass center, which is 200 mm from O_4. Link 4 is currently rotating at 1.186 rad/s clockwise with an angular acceleration of 12.51 rad/s² counter clockwise. Link 3 has a mass of 5.2 kg and a mass moment of inertia of 0.088 kg·m². Link 3 is currently horizontal ($\theta_3 = 180°$ or $0°$) and is rotating at 2.726 rad/s counter clockwise with an angular acceleration of 27.72 rad/s² counter clockwise. There is a 110 N weight centered 180 mm from pin B or 350 mm from pin A.

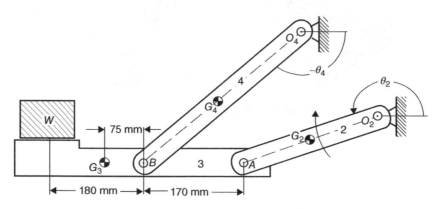

Figure 8.9   Problem 8.5

8.6   For the 4-bar linkage shown in Figure 8.10, determine the required input torque on link
      2 so that link 2 travels at a constant angular velocity of ω_2 equals 12 rad/s counter
      clockwise ($\alpha_2 = 0$). Link 2 is currently at θ_2 equals 68° with a length of 6 in. (L_2)
      and a weight of 1.61 lbs. Links 3 (L_3) and 4 (L_4) have a length of 12 in. Link 3 weighs
      3.22 lbs and link 4 has a weight of 4.83 lbs. Mass moments of inertia are $I_{G2} = 0.220$
      slug·in^2, $I_{G3} = 1.50$ slug·in^2, and $I_{G4} = 2.56$ slug·in^2. Fixed bearing O_4 is 15 in. directly
      right of fixed bearing O_2 (L_1). Analysis indicates that $\theta_3 = 31°$ and $\theta_4 = 101.9°$. Also,
      $\omega_3 = 3.539$ rad/s clockwise and $\omega_4 = 3.822$ rad/s counter clockwise, along with $\alpha_3 =$
      52.17 rad/s^2 counter clockwise and $\alpha_4 = 69.05$ rad/s^2 counter clockwise. Link 2's center
      of mass is located 2 in. from O_2 (r_2). What are the x and y components of G_2's accel-
      eration? Link 3's center of mass is located 4 in. from pin A (r_3). What are the x and y
      components of G_3's acceleration? Link 4's center of mass is located 3 in. from O_4 (r_4).
      What are the x and y components of G_4's acceleration? What are the magnitudes of the
      pin forces at O_2, A, B, and O_4?

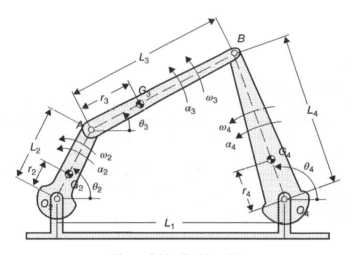

Figure 8.10   Problem 8.6

9

Spur Gears

9.1 Introduction

Gears are used to transmit motion from one rotating shaft to another shaft or from a rotating shaft to a body which translates. The translating body can be considered as rotating about an axis at infinity. In this chapter, we will only consider gears which provide a constant angular velocity ratio. Mating gears give the same motion to the shafts as a pair of equivalent rolling cylinders (see Figure 9.1).

The pinion is the name given to the smaller of the two mating gears. The larger gear is often referenced to as the gear. In this chapter, we will discuss only the case where the teeth are straight and parallel to the axis of rotation of the two shafts. These gears are called spur gears. A pair of spur gears is shown in Figure 9.2.

9.2 Other Types of Gears

Although only spur gears are going to be discussed in this chapter, it is important to know that other types of gears exist. The helical gear is like the spur gear except its teeth are at an angle known as the helix angle. This provides greater surface contact; thus, larger torques can be transmitted. However, helical gears create an axial force on the shaft. The herringbone and the double-helical gears do not transfer an axial force to the shaft as the helical gears do. Bevel gears allow power to be transmitted between shafts whose centerlines intersect, and their shafts are at an angle to each other. With hypoid gears, the shaft's centerlines do not intersect. Worm gears allow for a large gear reduction in a small space; however, they have low efficiencies due to high frictional forces (see Figure 9.3).

Design and Analysis of Mechanisms: A Planar Approach, First Edition. Michael J. Rider.
© 2015 John Wiley & Sons, Ltd. Published 2015 by John Wiley & Sons, Ltd.

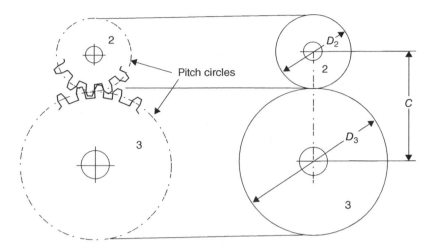

Figure 9.1   Gears and rolling cylinders

Figure 9.2   External meshing spur gears

9.3   Fundamental Law of Gearing

In Figure 9.4, a direct-contact mechanism is shown, and line L–L, a common normal to the contacting surfaces, interacts with the line of centers O_2O_3 at point P_P. It is shown that the angular velocity ratio for gears two and three is inversely proportional to the distances to their equivalent pitch circle. Note that gears 2 and 3 rotate in opposite directions:

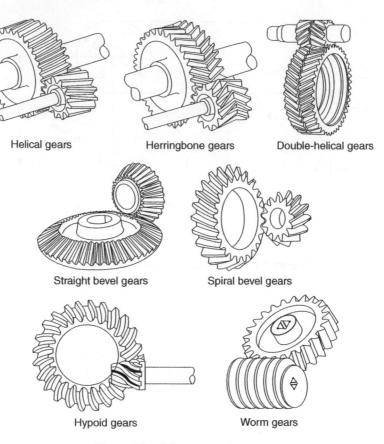

Helical gears   Herringbone gears   Double-helical gears

Straight bevel gears   Spiral bevel gears

Hypoid gears   Worm gears

Figure 9.3   Other types of gears

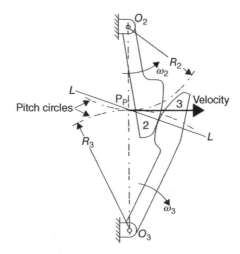

Figure 9.4   Line of action

Velocity of point $P_P = \omega_2 \cdot R_2 = \omega_3 R_3$

$$\left(\frac{\omega_2}{\omega_3}\right) = \left(\frac{R_3}{R_2}\right)$$

The action of a single pair of mating gear teeth as they pass through their entire phase of action must be such that the ratio of the angular velocity of the driven gear to that of the driving gear must remain constant. This is the fundamental criterion that governs the choice of tooth profiles. If this was not true, serious vibration and impact problems will result even at low speeds. The motion of gears 2 and 3 will then be equivalent to two imaginary rolling circles which are in contact of point P_P. These circles are known as pitch circles. The fundamental law of gearing for circular gears states that in order for a pair of gears to transmit a constant angular velocity ratio, the shape of their contacting profile must be such that the common normal passes through a fixed point on the line of centers. This is point P_P, the pitch point.

The involute curve is the path generated by a trace point on a cord as the cord is unwrapped from the base circle. This is shown in Figure 9.5 where "I" is the trace point for the involute curve. Note that the cord A–I is normal to the involute at I and the distance A–I is the instantaneous value of the radius of curvature. As the involute is generated from its origin I_0 to I_1, the radius of curvature increases continuously. It is zero at I_0 and increases as it moves to I_1.

If the two mating tooth profiles both have the shape of involute curves, the condition that the pitch point P_P remains stationary is satisfied (see Figure 9.4). Involute curves are conjugate curves that satisfy the fundamental law of gearing.

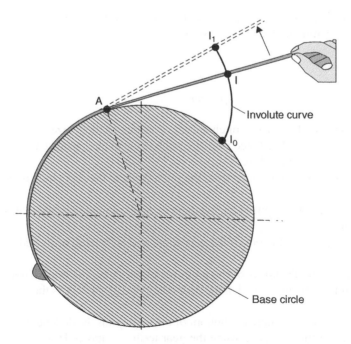

Figure 9.5   Involute curve generation

9.4   Nomenclature

Spur gears have pitch surfaces which are cylindrical in form. Their teeth are straight and parallel to the axis of rotation. Spur gears are the simplest type of gears, and hence, we will investigate them here. Many of the definitions and nomenclature for spur gears are the same for all types of gears.

The **pitch diameter**, D, is the diameter of the pitch circle. The pitch circle is a theoretical circle upon which all computations are made.

Pressure angle, ϕ, is the angle of contact between two mating teeth, typically 14.5°, 20°, or 25°.

The **circular pitch**, p, is the distance from a point on one tooth to the corresponding point on the next tooth measured along the pitch circle.

The **diametric pitch**, P_d (sometimes referred to as **pitch**), is used with English units and is the ratio of the number of teeth on a gear to the pitch circle diameter in inches. Let N be the number of teeth and D the diameter of the pitch circle in inches; then

$$P_d = \frac{N}{D}\left\{\frac{\text{teeth}}{\text{in}}\right\} \qquad (9.1)$$

The **module**, m, is used with SI units and is the ratio of the pitch circle diameter of the gear in millimeters to the number of teeth. Then

$$m = \frac{D}{N}\left\{\frac{\text{mm}}{\text{tooth}}\right\} \qquad (9.2)$$

Since circular pitch, p, is the circumference divided by the number of teeth, circular pitch is also equal to pi times the module for the SI system:

$$p = \frac{\pi D}{N} \qquad (9.3)$$

$$p = \pi m \qquad (9.4)$$

Circular pitch, module, and diametric pitch are all measures of tooth size.

The **addendum**, a, is the radial distance from the pitch circle to the outside circle of the gear or the addendum circle (see Figure 9.6).

The **dedendum**, b, is the radial distance from the pitch circle to the root circle or the dedendum circle.

The **working depth**, h_w, is the depth of engagement of a pair of mating gears; that is, it is the sum of their addendums.

The **whole depth**, h_t, is the full depth of the gear tooth and is the sum of its addendum and dedendum.

The **tooth thickness** is the thickness of the tooth measured along the pitch circle.

The **clearance**, c, is the amount by which the dedendum of a gear exceeds the addendum of the mating gear.

The **fillet** is the concave curve where the bottom of the tooth joins the dedendum circle. The fillet radius, r_f, is equal to the clearance when the gear teeth are drawn. However, the actual shape of the fillet curve on a gear will depend upon the method used for cutting the teeth.

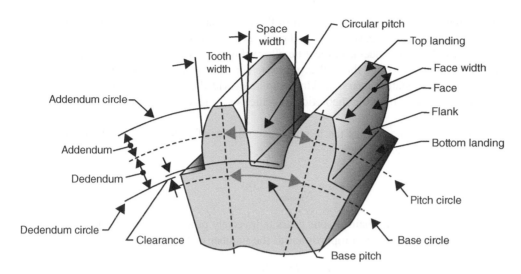

Figure 9.6   Tooth profile

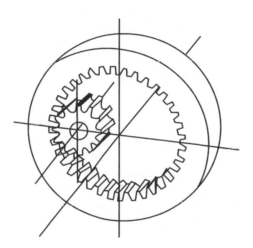

Figure 9.7   Internally meshing spur gears

The **pinion** is the smaller of the two mating gears.

The **gear** is the larger of the two mating gears.

The **center distance**, C, is the distance from center to center for two externally mating gears:

$$C = \frac{D_2 + D_3}{2} = \frac{D_{\text{pinion}} + D_{\text{gear}}}{2}$$

The **center distance**, C, is the distance from center to center for two internally mating gears (see Figure 9.7):

$$C = \frac{D_3 - D_2}{2} = \frac{D_{\text{gear}} - D_{\text{pinion}}}{2}$$

Backlash is the amount by which the width of the tooth space on a gear exceeds the tooth thickness on the mating gear, measured along the pitch circles. Theoretically, backlash should be zero, but in practice, some backlash must be allowed to prevent jamming of the gear teeth due to tooth machining errors and thermal expansion. Throughout this text, zero backlash will be assumed. For a set of mating gears, backlash can be provided either by mounting the gears at a center distance greater than the theoretical distance or by feeding the tooth cutter deeper than normal.

The **gear ratio**, G_R, is the ratio of the larger gear's number of teeth to the smaller pinion's number of teeth for two mating gears:

$$G_R = \frac{N_{\text{gear}}}{N_{\text{pinion}}} \geq 1.00 \tag{9.5}$$

Angular velocity ratio for a pair of spur gears is inversely proportional to the diameter of their pitch circles and/or inversely proportional to the number of teeth:

$$\frac{\omega_2}{\omega_3} = \frac{D_3}{D_2} = \frac{N_3}{N_2} \tag{9.6}$$

Hobbing is a common, high-precision method used to cut gear teeth. Check these out on YouTube: https://www.youtube.com/watch?v=zt7GfAt02I0

Broaching/shaping is used to create internal gears and racks. Check these out on YouTube: https://www.youtube.com/watch?v=fmTSXNE8OVw

https://www.youtube.com/watch?v=gPZ33HTcEdo

9.5 Tooth System

The tooth system is the name given to a standard that specifies the relationship between addendum, dedendum, clearance, tooth thickness, and fillet radius to attain interchangeability of gears of all tooth values having the same pressure angle and diametric pitch or module. In order for a pair of spur gears to mesh properly, they must have the same pressure angle and the same tooth size as specified by their diametric pitch or module. The number of teeth and the pitch diameters of the two gears need not match, but are chosen to give the desired speed ratio.

The size of the teeth is chosen by selecting the diametric pitch or module. Standard cutters are generally available for the sizes listed in Table 9.1.

Table 9.1 Standard tooth sizes

Standard diametric pitches, US customary (teeth/in.)	
Course	1, 1¼, 1½, 1¾, 2, 2¼, 2½, 2¾, 3, 3½, 4, 5, 6, 8, 10, 12, 16
Fine	20, 24, 32, 40, 48, 64, 80, 96, 120, 150, 200
Standard modules, SI units (mm/tooth)	
Preferred	0.5, 1, 1.5, 2, 2.5, 3, 4, 5, 6, 8, 10, 12, 16, 20, 25, 32, 40
Acceptable	1.25, 1.75, 2.25, 2.75, 3.5, 4.5, 5.5, 6.5, 7, 9, 11, 14, 18

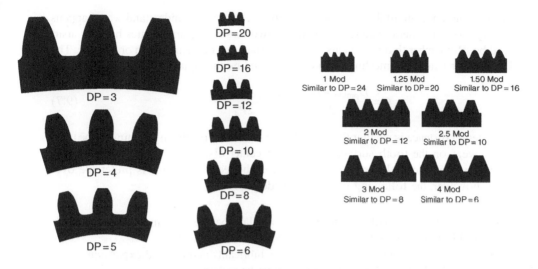

Figure 9.8 Diametric pitches and module sizes

Table 9.2 Standard tooth system for spur gears

Tooth system	Pressure angle, ϕ	Addendum, a	Dedendum, b
Full depth	14.5°	$1/P_d$ or 1 m	$1.157/P_d$ or 1.157 m
Full depth	20 or 25°	$1/P_d$ or 1 m	$1.25/P_d$ or 1.25 m
Stub teeth	20°	$0.8/P_d$ or 0.8 m	$1.00/P_d$ or 1.00 m

Table 9.3 Preferred number of teeth on spur gears

Standard number of teeth (common integers)

Typical stock	10–26, 28, 30, 32, 33, 35, 36, 39, 40, 42, 44, 45, 48, 52, 54, 56, 60, 64, 66, 68, 70, 72, 75, 76, 80, 84, 88, 90, 92, 96, 100, 105, 108, 110, 112, 120, 128, 130, 132, 144, 160, 168, 180

Figure 9.8 shows the relative size between some diametric pitches and some modules. Note that as the diametric pitch increases, the tooth size decreases. Also, note that as the modules sizes increases, the tooth size increases. Once the diametric pitch or module has been chosen, the remaining dimensions of the tooth are set by the standard shown in Tables 9.2. Table 9.3 shows the preferred number of teeth on commercially available spur gears.

9.6 Meshing Gears

It is important to properly set up a set of meshing gears to ensure the system will be both durable and accurate. There is an ideal distance between two meshing gears. At this distance, tooth contact is continuous and smooth. Since this perfect center is impossible to achieve due to manufacturing tolerances in all the parts of the assembly, we have to understand what happens when we vary from this ideal distance. The size of two meshing gears dictates how far apart they should be. Remember that both gears must have the same diametric pitch or module. The center distance for externally meshing gears is governed by the equation

$$C = \frac{D_{\text{pinion}} + D_{\text{gear}}}{2} = \frac{N_{\text{pinion}} + N_{\text{gear}}}{2P_d} \tag{9.7}$$

When the center distance is smaller than the ideal, interference results in jamming of the teeth or excessive wear on the tooth surfaces. When the center distance is larger than the ideal distance, the result is backlash, or clearance between the teeth. Backlash is preferable over interference; however, the following must be considered:

- Changes to input direction will need to take up backlash before affecting the output.
- Excessive backlash will disrupt the smooth engagement of the teeth.
- Clearance from backlash can be advantageous for lubricants and thermal expansion.
- The pressure angle changes; thus, interference must be rechecked.

9.6.1 Operating Pressure Angle

The actual operating pressure angles change any time the center distance deviates from the ideal center distance. The new operating pressure angle can be calculated using the following equation:

$$\phi_{\text{operating}} = \cos^{-1}\left(\frac{C_{\text{ideal}}}{C_{\text{operating}}} \cos\phi\right) \tag{9.8}$$

9.6.2 Contact Ratio

The contact ratio, CR, is the average number of teeth in contact at any given instant. This value should be between 1.4 and 2.0. With a contact ratio of 1.0, one tooth is just leaving the contact area as another tooth enters the contact area. This is undesirable because it can cause vibrations, noise, and variations in the velocity ratio. Also, the applied load will be applied to the tip of the tooth, creating a large bending moment. This equation assumes the addendum is $1/P_d$ and no interference occurs:

$$CR = \frac{\sqrt{(N_p + 2)^2 - (N_p \cos\phi)^2} + \sqrt{(N_g + 2)^2 - (N_g \cos\phi)^2} - (N_p + N_g)\sin\phi}{2\pi\cos\phi} \tag{9.9}$$

where
N_p = number of teeth on the pinion
N_g = number of teeth on the gear
ϕ = pressure angle

Contact ratios greater than 1 allow load sharing between two teeth. A contact ratio of 1.7 means that 70% of the time two teeth are sharing the applied load and 30% of the time the load is carried by only one tooth.

Example 9.1   Calculate the contact ratio for two mating gears

Problem: Calculate the contact ratio for an 18-tooth pinion meshing with a 30-tooth gear. The diametric pitch is 8 teeth/in. and the pressure angle is 20°:

$$CR = \frac{\sqrt{(N_p+2)^2 - (N_p\cos\phi)^2} + \sqrt{(N_g+2)^2 - (N_g\cos\phi)^2} - (N_p+N_g)\sin\phi}{2\pi\cos\phi}$$

$$CR = \frac{\sqrt{(18+2)^2 - (18\cos20°)^2} + \sqrt{(30+2)^2 - (30\cos20°)^2} - (18+30)\sin20°}{2\pi\cos20°}$$

$$CR = \frac{9.397}{5.904} = 1.592$$

Answer: $\boxed{CR = 1.59}$ or 59% of the time two teeth are sharing the applied load.

9.7   Noninterference of Gear Teeth

Involute gear teeth have involute profiles between the base circle and the addendum circle. If the dedendum circle lies inside the base circle, then the portion of the tooth profile between the base circle and the dedendum circle will not be an involute curve. Thus, if the tip of the mating gear makes contact below the base circle, interference occurs and the fundamental law of gearing does not apply. That is, the velocity ratio will not remain constant.

If the addendum circle of a gear intersects the line of action outside the tangent point between the base circle and the line of action of the mating gear, then interference occurs. Interference of the pinion teeth needs to be checked against the larger gear. If there is no interference here, the large gear's teeth will not interfere with the pinion (see Figure 9.9):

$$R_{pp} = \frac{N_p}{2P_d} \quad \text{and} \quad R_{pb} = R_{pp}\cos\phi$$

$$R_{gp} = \frac{N_g}{2P_d} \quad \text{and} \quad R_{gb} = R_{gp}\cos\phi$$

$$C = R_{pp} + R_{gp}$$

$$R_{pa\_max} = \sqrt{R_{gb}^2 + C^2\sin^2\phi}$$

$$a_p = \frac{1}{P_d} \quad \text{(For standard tooth system)}$$

$$R_{pa} = R_{pp} + a_p$$

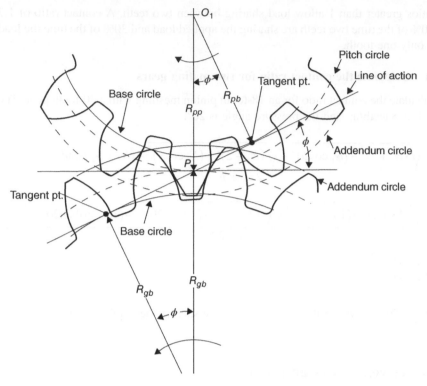

Figure 9.9   Check for interference

where

R_{pp} = radius of the pinion's pitch circle
R_{pb} = radius of the pinion's base circle
R_{gp} = radius of the gear's pitch circle
R_{gb} = radius of the gear's base circle
C = center distance between the two mating gears
R_{pa_max} = maximum allowable pinion addendum radius
a_p = size of the addendum on pinion
R_{pa} = actual pinion addendum radius

If ($R_{pa} < R_{pa\_max}$), then no interference occurs. If the gear teeth are shortened or the pressure angle is varied, make the appropriate adjustments, and then check for interference.

For all of the equations that follow, remember that these equations are for full-depth standard gear teeth, that $N_p \leq N_g$, and that $G_R = N_g/N_p \geq 1$.

To avoid interference, the following conditions are necessary for standard teeth. The minimum pinion size for a given gear ratio, G_R, is

$$N_p \geq \frac{2\left(G_R + \sqrt{G_R{}^2 + (1 + 2G_R)\sin^2\phi}\right)}{(1 + 2G_R)\sin^2\phi} \tag{9.10}$$

For a G_R of 1.0 (equal size gears), the smallest pinion size, N_p, is shown in Table 9.4. The minimum pinion size for a gear rack is

$$N_p \geq \frac{2}{\sin^2 \phi} \qquad (9.11)$$

For a G_R of infinity (mesh with a gear rack), the smallest pinion size is shown in Table 9.5.

The results above for the minimum number of teeth for a pinion to mesh with a rack show a major disadvantage of the 14.5° pressure angle gear system. The gears must be larger in size for smaller pressure angles to avoid interference with standard size teeth.

The maximum gear size for a given pinion size, N_p, is (see Table 9.6)

$$N_g \leq \frac{\left(N_p \sin \phi\right)^2 - 4}{4 - 2N_p \sin^2 \phi} \qquad (9.12)$$

Table 9.4 Minimum pinion size without interference or undercutting

Pressure angle, ϕ	$N_{p\,min}$ (teeth)
14.5°	23
20°	13
25°	9

Table 9.5 Minimum pinion size for a gear rack

Pressure angle, ϕ	$N_{p\,min}$ (teeth)
14.5°	32
20°	18
25°	12

Table 9.6 Maximum gear size without interference or undercutting

Pressure angle, ϕ	N_p (teeth)	$N_{g\,max}$ (teeth)
14.5°	23	26
14.5°	24	32
14.5°	25	40
14.5°	26	51
14.5°	27	67
14.5°	28	92
14.5°	29	133
14.5°	30	219
14.5°	31	496
14.5°	32	Rack

If the value for N_g comes out negative, then the pinion will mesh with a rack (N_g = infinity). The above equation can be used to create a similar table for 20 and 25° pressure angles.

Example 9.2   Determine the operating pressure angle due to incorrect center distance between two mating gears

Problem: Determine the operating pressure angle due to the center distance between two mating gears of 25 teeth and 35 teeth with a pressure angle of 14.5° and a diametric pitch of 10 teeth/in. The operating center distance is measured at 3.246 in. Is there now interference?

Solution: Calculate the new operating pressure angle, and then check for interference using standard gear teeth. The ideal center distance is 3.250 in.:

$$\phi_{operating} = \cos^{-1}\left(\frac{C_{ideal}}{C_{operating}}\cos\phi\right) = \cos^{-1}\left(\frac{3.250}{3.246}\cos 14.5°\right) = 14.2°$$

$$N_g \le \frac{(N_p\sin\phi)^2 - 4}{4 - 2N_p\sin^2\phi} = \frac{(25\sin(14.2°))^2 - 4}{4 - 2(25)\sin^2(14.2°)} = \frac{33.61}{0.991} = 33.9$$

$N_g \le 33$ teeth without interference

Answer: | Interference is present now due to center distance variation of 0.004 in. |

The pitch line velocity, v_t, in feet per minute is the linear speed of a tooth at the pitch circle radius:

$$v_t(\text{f.p.m.}) = \frac{\pi Nn}{12P_d} = \frac{\pi mN}{304.8} \tag{9.13}$$

where
N = number of teeth
n = speed of gear in revolutions per minute (r.p.m.)
P_d = diametric pitch
m = module in SI system

9.8   Gear Racks

A basic understanding of gear rack systems is useful when dealing with the products that use them, like automotive rack and pinion steering, linear actuators and encoders, and adjustable optics in microscopes (see Figure 9.10). As with other gear types, meshing gears and racks must have the same diametric pitch. The number of gear teeth advanced radially on the pinion equals the number of rack teeth advanced linearly on the rack. The arc length between gear teeth at the pitch diameter is the same as the length of the line segment between rack teeth at the pitch line:

Linear pitch of the rack = circular pitch of the pinion

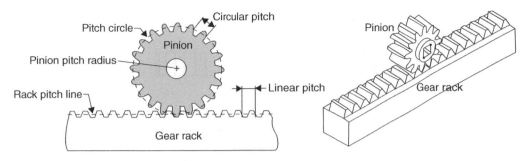

Figure 9.10   Gear rack and pinion

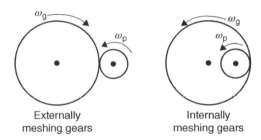

Externally
meshing gears

Internally
meshing gears

Figure 9.11   External and internal meshing

At this tangency between the pitch diameter of the gear and pitch line of the rack, rolling without slipping is an accurate model.

9.9   Gear Trains

Internal combustion engines and electric motors run efficiently and produce maximum power at high speeds, while mechanisms and other machinery tend to operate at lower speeds. Rolling surfaces fill this need for speed reduction. However, the rolling surfaces are replaced by toothed gears to eliminate slipping and increase the overall efficiency of the system. Spur gears transmit motion and torque from one parallel shaft to another. Over 2000 years ago, the Greeks and the Chinese used spoked gears to transmit power from rotating water wheels. Design changes have led to gears with the positive engagement of their teeth and the ability for continuous rotation while transmitting power.

A simple gear train is one with only one gear per shaft and each shaft rotating about a fixed axis. A compound gear train is one where two or more gears are fastened together and rotate about a common shaft.

The gear ratio for externally meshing gears are considered to be negative since the two gears will rotate in opposite directions. The gear ratio for internally meshing gears is considered to be positive since both gears will rotate in the same direction (see Figure 9.11).

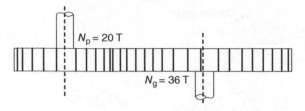

Figure 9.12   Externally meshing spur gears

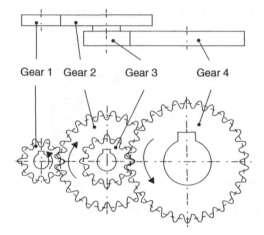

Figure 9.13   Compound gear train

9.9.1   Simple Gear Train

The input shaft is rotating at 500 r.p.m. clockwise (c.w.) and drives a 20-tooth spur gear. The output shaft is attached to a 36-tooth spur gear with a diametric pitch of 8 teeth/in. and a pressure angle of 20°. What are the output speed and the ideal center distance (see Figure 9.12)?

$$\omega_{\text{gear}} = \omega_{\text{pinion}}\left(-\frac{N_{\text{pinion}}}{N_{\text{gear}}}\right) = 500\left(-\frac{20}{36}\right) = -278 = 278\,\text{r.p.m. c.c.w}$$

$$C = \frac{N_{\text{pinion}} + N_{\text{gear}}}{2P_d} = \frac{20 + 36}{2(8)} = 3.50''$$

   Remember that externally meshing spur gears reverse the direction of rotation of the output shaft.

9.9.2   Compound Gear Train

A compound gear train consists of two or more sets of mating gears as shown in Figure 9.13. The center distance between the input shaft and the output shaft can be calculated by calculating the center distance for each pair of mating gears and then adding them:

$$\frac{\omega_{\text{out}}}{\omega_{\text{in}}} = \left(-\frac{N_1}{N_2}\right)\left(-\frac{N_3}{N_4}\right) = +\frac{N_1 N_3}{N_2 N_4}$$

$$C = \frac{N_1 + N_2}{2P_{d12}} + \frac{N_3 + N_4}{2P_{d34}}$$

where

P_{d12} = diametric pitch for gears 1 and 2
P_{d34} = diametric pitch for gears 3 and 4

Example 9.3   Design a compound gear train with a specified gear ratio and center distance

Design a planar gear train containing two sets of mating spur gears with an overall gear ratio of 18:1. Each set of mating gears must have a gear ratio between 3 and 6 exclusive. The center distance between the input shaft and the output shaft must be approximately 8.00 in. All spur gears have a 20° pressure angle. Acceptable diametric pitches for all spur gears are 14, 12, and 10 teeth/in. No gear tooth interference should be present in the final design, that is, $N_P > 15$ teeth. No gear may be larger than 160 teeth. (Note that $\sqrt{18} = 4.24$.)

The procedure needs to search through gear ratios greater than 3 and less than 4.25. The second pair of mating spur gears will then have a gear ratio greater than 4.24 and less than 6. After a solution is found, the first and second set of mating gears may be switched without affecting the overall gear ratio of 18–1 or the center distance near 8.00 in.

Procedure:

1. Start with pinion #1 having 16 teeth and increment it to a size of 21 teeth. You may go higher, but this will keep the search range small.
2. Note that the gear ratio of the first two gears times the gear ratio of the second two gears must have a product of 18:

$$G_{R12} * G_{R34} = G_{R\text{total}} = 18$$

3. Select a gear ratio greater than 3.00, and then calculate the number of teeth on the mating gear. Try to get the calculated number of teeth to be an integer since the number of teeth must be an integer:

$$N_2 = G_{R12} * N_1$$

Or select the number of teeth for gear two so that the gear ratio is greater than 3.00:

$$G_{R12} = \frac{N_2}{N_1}$$

4. Start with pinion #3 having 16 teeth and increment it to a size of 21 teeth. You may go higher, but this will keep the search range small.

5. Calculate the number of teeth on gear 4 so that the overall gear ratio is 18:1. Try to get the calculated number of teeth to be an integer since the number of teeth must be an integer:

$$N_{4\text{calculate}} = \frac{18N_3}{G_{R12}}$$

6. Round the calculated number of teeth on gear 4 to the nearest integer, and then use it to calculate the overall gear ratio and compare it with the desired ratio of 18–1:

$$G_{R\text{total}} = \frac{N_2 N_4}{N_1 N_3}$$

7. Once you have an overall gear ratio of 18 or very close to 18, calculate the center distance between the input shaft and the output shaft:

$$C = \frac{N_2 + N_1}{2P_{d12}} + \frac{N_4 + N_3}{2P_{d34}}$$

8. Select the diametric pitch for gears 1 and 2 along with the diametric pitch for gears 3 and 4 so that the center distance gets as close as possible to the specified center distance.
9. Proceed until a solution is found or a solution close enough to the ideal solution is found. This depends upon whether the overall gear ratio and/or the center distance must be exact or just close. This entire procedure can easily be done using a spreadsheet program.

   A partial solution for this gear train design follows.

N1	N2	GR12	N3	N4	N4_ calc	GR34_ calc	GR_ total	Pd12	Pd34	Ctr Dist	Comment
16	49	3.063	17	100	99.92	5.8776	18.01	10	12	8.125	
17	52	3.059	16	94	94.15	5.8846	17.97	10	12	8.033	
18	55	3.056	18	106	106.04	5.8909	17.99	10	14	8.079	
19	58	3.053	19	112	112.03	5.8966	17.99	12	14	7.887	
20	61	3.05	20	118	118.03	5.9016	18.00	14	14	7.821	
21	64	3.048	21	124	124.03	5.9063	18.00	14	14	8.214	$C > 8.00''$
20	62	3.1	20	116	116.13	5.8065	17.98	14	14	7.786	
16	50	3.125	20	115	115.20	5.7600	17.97	14	12	7.982	Close
20	74	3.7	22	107	107.03	4.8649	18.00	14	14	7.964	better

From the partial solution shown above, the last entry appears to be the best solution so far:
$\boxed{N_1 = 20\,\text{T}}$, $\boxed{N_2 = 74\,\text{T}}$, $\boxed{N_3 = 22\,\text{T}}$, $\boxed{N_4 = 107\,\text{T}}$, $\boxed{GR = 18.00 = 3.700 * 4.864}$, $\boxed{C = 7.964\,\text{in}}$.

A search further shows that there is a better solution than the one shown here. The question is, how important is it to see if a solution with a gear ratio of 18–1 and a center distance of exactly 8.00 in. can be found? Is the solution above close enough?

9.9.3   Inverted Compound Gear Train

The inverted compound gear train is such that the input shaft and the output shafts are along the same centerline as shown in Figure 9.14. The speed ratio and the necessary gear teeth selection are shown below:

$$\frac{\omega_{out}}{\omega_{in}} = \left(-\frac{N_1}{N_2}\right)\left(-\frac{N_3}{N_4}\right) = +\frac{N_1 N_3}{N_2 N_4}$$

$$C = \frac{N_1 + N_2}{2P_{d12}} = \frac{N_3 + N_4}{2P_{d34}}$$

where

P_{d12} = diametric pitch for gears 1 and 2
P_{d34} = diametric pitch for gears 3 and 4

Note that ω_{out} and ω_{in} rotate in the same direction.

Example 9.4   Design an inverted compound gear train with a specified gear ratio

Problem: Design an inverted compound gear train with a gear ratio of 40:1. All gears have a 20° pressure angle. The preferred center distance is 5.00 in.

Solution: Set up an MS Excel spreadsheet, and then adjust the numbers to obtain a feasible solution. The tooth size for the second set of gears must be larger than or equal to that of the first set of gears:

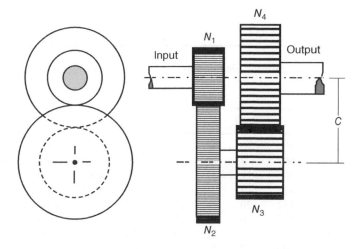

Figure 9.14   Inverted compound gear train

$$G_R = 40 = \frac{N_2 N_4}{N_1 N_3}$$

$$C = \frac{N_1 + N_2}{2 P_{d12}} = 5.00'' = \frac{N_3 + N_4}{2 P_{d34}}$$

$$P_{d12} \geq P_{d34}$$

Above are three equations and an inequality with six unknowns. For a 20° pressure angle, the number of teeth on the pinion must be greater than or equal to 16 to avoid interference. (16T meshes with up to 101T leading to a G_R of 6.31, and $6.31^2 = 39.8 < 40$.) Common sense plays a large part in how long it might take you to find a solution.

Procedure:

1. Pick N_1 and P_{d12}.
2. Calculate N_2 and verify that $(4.5 < G_{R12} < 9)$:

$$N_2 = 2 P_{d12}(5.00) - N_1$$

3. Pick P_{d34}.
4. Calculate N_3 and N_4 and verify that $(4.5 < G_{R34} < 9)$:

$$N_3 = \frac{2 P_{d34}(5.00) N_2}{40 N_1 + N_2}$$

$$N_4 = 2 P_{d34}(5.00) - N_3$$

Worksheet using equations above.

Pd12	N1	N2	GR12	Pd34	N3	N4	GR34
20	**31**	169	5.452	**20**	23.989	176.011	7.337

Here are some of the possible solutions obtained by the trial-and-error method after rounding N_3 and N_4 to integers.

Pd12	N1	N2	GR12	Pd34	N3	N4	GR34	GR	C12	C34
12	16	104	6.500	10	14	86	6.143	39.93	5.000	5.000
12	18	102	5.667	8	10	70	7.000	39.67	5.000	5.000
16	16	144	9.000	12	22	98	4.455	40.09	5.000	5.000
16	17	143	8.412	12	21	99	4.714	39.66	5.000	5.000
20	31	169	5.452	20	24	176	7.333	39.98	5.000	5.000
20	26	174	6.692	16	23	137	5.957	39.86	5.000	5.000
18	19	161	8.474	12	21	99	4.714	39.95	5.000	5.000
18	20	160	8.000	12	20	100	5.000	40.00	5.000	5.000

Answers: $\boxed{N_1 = 20\text{T}, N_2 = 160\text{T}, P_{d12} = 18\text{T/in.}}$, $\boxed{N_3 = 20\text{T}, N_4 = 100\text{T}, P_{d34} = 12\text{T/in.}}$,
$\boxed{G_R = 40}$, and $\boxed{C = C_{12} = 5.000\,\text{in.} = C_{34}}$.

What if you obtain several valid solutions and you want to determine the best solution. First of all, you need to determine what makes the best design. Is it the smallest footprint? Is it the smallest rotating inertia? Is it the lightest design by weight?

9.9.4   Kinetic Energy of a Gear

Assume a gear is a solid rotating disk with its diameter equal to the pitch diameter so that its mass moment of inertia is ½m r^2 (see Figure 9.15):

$$I = \frac{1}{2}\text{mr}^2$$

$$m = \rho \cdot \text{volume}$$

$$\text{Volume} = \pi r^2 \cdot \text{thickness}$$

$$r = \frac{D}{2} = \frac{N}{2P_d}$$

or

$$I = \frac{\rho \pi r^4 \cdot \text{thickness}}{32} = \frac{\rho \pi \cdot \text{thickness}}{32}\left(\frac{N}{P_d}\right)^4$$

And the kinetic energy for a rotating gear is

$$K.E. = \frac{1}{2}I\omega^2 = \frac{\rho\pi \cdot \text{thickness}}{64}\left(\frac{N}{P_d}\right)^4 \omega^2 \qquad (9.14)$$

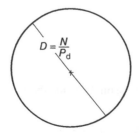

Figure 9.15   Gear as solid disk

Example 9.5   Minimized the rotating inertia (or kinetic energy) of a compound gear system

Problem: Design an 18:1 compound gear train with an overall center distance between the input shaft and the output shaft of approximately 8 in. The diametric pitch must be the same for all gears, and it must be less than or equal to 14 (14, 12, 10, 8). The first pinion in the gear train must have 20 teeth and is 1.00 in. thick. All gears are made from steel and have the same thickness. The best design is the one with the smallest rotating inertia. Ignore the rotating inertia of the shafts and bearings. The input gear is rotating clockwise at 180 rad/s.

Solution: A method similar to the one used previously was used to obtain these solutions; however, the center distance was left to chance since it only had to be near 8.00 in. The diametric pitch was selected along with three of the gears. The fourth was calculated so that the overall gear ratio was essentially 18.

Below are six possible solutions with a gear ratio at or near 18:1 and a center distance of approximately 8.00 in.

P_d (T/in.)	N_1	N_2	N_3	N_4	G_R	C (in.)
14	20	62	20	116	17.98	7.786
14	20	66	22	120	18.00	8.143
14	20	70	21	108	18.00	7.821
14	20	84	21	90	18.00	7.679
12	20	64	16	90	18.00	7.917
12	20	68	17	90	18.00	8.125

To determine the best solution, we need to calculate the kinetic energy of each gear set, and then select the lowest kinetic energy gear set. The total kinetic energy of the gear set is the sum of all its components:

$$K.E. = K.E._1 + K.E._2 + K.E._3 + K.E._4$$

$$K.E. = \frac{\rho \pi \cdot \text{thickness}}{64} \left[\left(\frac{N_1}{P_{d12}}\right)^4 \omega_1^2 + \left(\frac{N_2}{P_{d12}}\right)^4 \omega_2^2 + \left(\frac{N_3}{P_{d34}}\right)^4 \omega_3^2 + \left(\frac{N_4}{P_{d34}}\right)^4 \omega_4^2 \right]$$

where

$$P_d = P_{d12} = P_{d34}$$

$$\omega_2 = \left(-\frac{N_1}{N_2} \right) \omega_1 = \omega_3 \text{ (gears 2 and 3 are on the same shaft)}$$

$$\omega_4 = \left(-\frac{N_3}{N_4} \right) \omega_3 = \left(-\frac{N_1}{N_2} \right) \left(-\frac{N_3}{N_4} \right) \omega_1 = \left(\frac{N_1 N_3}{N_2 N_4} \right) \omega_1$$

Simplifying

$$K.E. = \frac{\rho\pi \cdot \text{thickness}}{64 P_d{}^4}\left[N_1{}^4\omega_1{}^2 + N_2{}^4\omega_2{}^2 + N_3{}^4\omega_3{}^2 + N_4{}^4\omega_4{}^2\right]$$

$$K.E. = \frac{\rho\pi \cdot \text{thickness}}{64 P_d{}^4}\left[N_1{}^4\omega_1{}^2 + N_2{}^4\left(-\frac{N_1}{N_2}\right)^2\omega_1{}^2 + N_3{}^4\left(-\frac{N_1}{N_2}\right)^2\omega_1{}^2 + N_4{}^4\left(\frac{N_1 N_3}{N_2 N_4}\right)^2\omega_1{}^2\right]$$

$$K.E. = \frac{\rho\pi \cdot \text{thickness}}{64 P_d{}^4}\left[N_1{}^4 + N_2{}^2 N_1{}^2 + N_3{}^4\left(\frac{N_1}{N_2}\right)^2 + N_4{}^2\left(\frac{N_1 N_3}{N_2}\right)^2\right]\omega_1{}^2$$

Note that to minimize the total kinetic energy of the gear set, only the following terms need to be analyzed since everything else is a constant:

$$\text{Minimize}\left[N_1{}^4 + N_2{}^2 N_1{}^2 + N_3{}^4\left(\frac{N_1}{N_2}\right)^2 + N_4{}^2\left(\frac{N_1 N_3}{N_2}\right)^2\right]$$

N1	N2	N3	N4	C1*KE
20	62	20	116	7.5382E+09
20	66	22	120	9.2179E+09
20	70	21	108	4.8999E+09
20	84	21	90	1.6432E+09
20	64	16	90	1.6421E+09
20	68	17	90	1.6423E+09

Answer: $\boxed{N_1 = 20\text{T}, N_2 = 64\text{T}, N_3 = 16\text{T}, N_4 = 90\text{T}}$, $\boxed{GR = 18.00}$, $\boxed{C = 7.917''}$, and $\boxed{\text{K.E. minimum}}$.

9.10 Planetary Gear Systems

In the previous section, ordinary fixed-axis gear trains were discussed. These are gear trains in which each gear rotates with respect to a center that is fixed to the frame. Ordinary gear trains have one degree of freedom. Turning the input gear forces the output gear to rotate.

A second kind of gear train is the planetary or elliptical gear train shown in Figure 9.16. A gear train in which the axis of one or more gears moves relative to the frame is referred to as a planetary or elliptical gear train. The gear at the center is called the sun since gears revolve around it. The outer gear is referred to as the ring gear. The gears between the ring gear and the sun gear are referred to as planet gears since they revolve around the sun gear. The distance between the center of the sun gear and the center of the planet gears is held constant by the arm which is also allowed to rotate.

Although a planetary gear train typically has three or four planet gears to evenly distribute the load and balance the device, for analysis purposes only one planet gear is shown as seen in Figure 9.17.

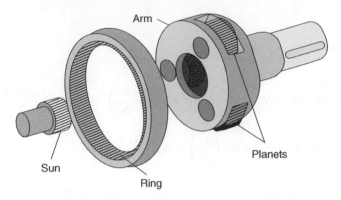

Figure 9.16   Standard planetary gear system

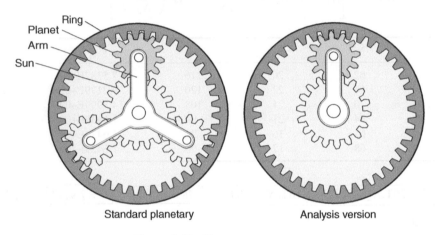

Standard planetary   Analysis version

Figure 9.17   Planetary gear system

In contrast to ordinary gear trains, a planetary or elliptical gear train can provide two degrees of freedom. It does this by releasing one of the gear centers from ground. A planetary gear system with two degrees of freedom can be used to combine two inputs into one output. Two motors can be combined to drive a single output shaft. For applications requiring only one degree of freedom, a planetary gear train can restrict one of its compounds to ground, thus removing one degree of freedom. This can be seen in Figure 9.18.

Planetary gear systems are more expensive to manufacture and maintain than ordinary fixed-axis gear trains; however, designers may choose them for three reasons. First, there is some situations in which two degrees of freedom are required. Second, when it comes to one degree of freedom power transmission from an input shaft to an output shaft, it is often possible to get the same fixed gear ratio within less space, and transmit more power, using planetary gear systems rather than ordinary gear systems. Third, planetary gear trains operate at a high efficiency just like fixed-axis gear trains.

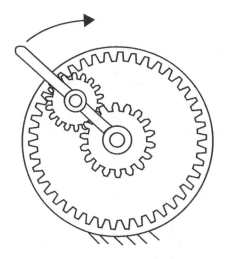

Figure 9.18   One degree of freedom

Figure 9.19   Vehicle differential

9.10.1   Differential

When a planetary gear train is allowed to maintain its two degrees of freedom, it is called a differential (see Figure 9.19). Differentials are useful when it is necessary to combine two inputs to produce one output. The differential on a vehicle allows for the two driving tires to rotate at different speeds when the vehicle is going around a curve. Without the two degrees of freedom, the two driving tires would be forced to rotate at the same speed, even though they are traveling different distances in the same amount of time. The result would be one drive tire skidding or slipping when the vehicle was turning a corner.

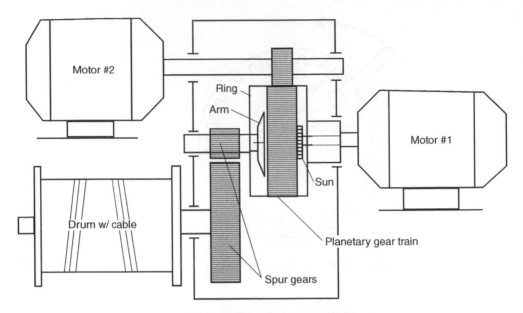

Figure 9.20   Two input motors driving drum

Differentials are also useful for combining the output of two motors. Cranes and hoists can be operated by using two motors to provide power. If one motor fails, the other motor continues to move the load at a reduced speed. In Figure 9.20, the right-hand motor drives the sun gear and the left-hand motor drives the ring gear. The arm is connected to the output shaft which winds the cable onto the drum after passing through a set of spur gears. With both motors running at the same speed and in the same direction, the arm rotates at the same speed and in the same direction. If one motor fails, the arm is driven by the remaining motor at half the original speed.

9.10.2   Clutch

As noted earlier, planetary gear systems inherently possess two degrees of freedom. By taking advantage of this fact, planetary gear trains may be used either to transmit power or to free-wheel, that is, rotate without transferring any power from the input shaft to the output shaft. This is illustrated in Figure 9.21. The sun gear is the driving gear and the arm is the driven member. When the brake is released, both the ring gear and the planetary gears are free to rotate, thus giving the system two degrees of freedom. No power will be transmitted from the input sun gear to the output arm. When the ring gear brake is applied, thus stopping the ring gear from rotating, the system has only one degree of freedom. As the sun gear rotates, both the planet gears and the arm must rotate, and power will be transmitted through the arm to the output shaft.

9.10.3   Transmission

If the gear train is to transmit power from a single input shaft to a single output shaft, then the gear train must have only one degree of freedom. Such a gear train is often referred to as a transmission, because it transmits power from the input to the output. The planetary gear system may function as a transmission by fixing one member to ground, thus removing one degree of freedom.

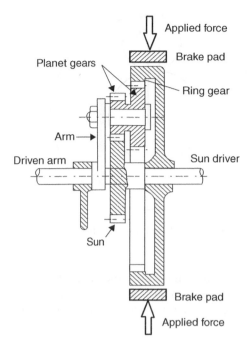

Figure 9.21   Planetary clutch

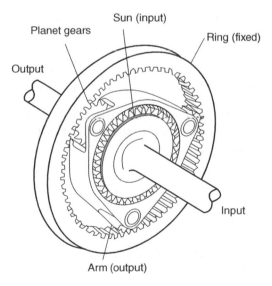

Figure 9.22   Planetary creeper drive

The "creeper drive" transmission shown in Figure 9.22 illustrates this application. Creeper drive is an option on some tractors or trucks. It provides an additional speed reduction. It is used when the tractor or truck needs to move very slowly or supply extra torque to its drive wheels. The internal ring gear is press fit into the housing so that it will not rotate, thus

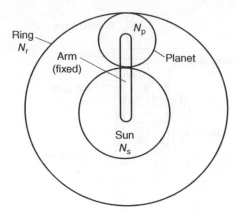

Figure 9.23   Fixed arm gear train

removing one degree of freedom. As the engine turns the sun gear, the planets and the arm are forced to orbit around the sun gear. The arm turns at a slower speed than the sun gear. In this example, the sun gear is the driver, the arm is the driven, and the planetary gear system acts as a speed reducer.

9.10.4   Formula Method

If the arm is held fixed, then the following planetary gear train behaves like the standard fixed-axis gear train (see Figure 9.23):

$$\frac{\omega_{\text{ring}}}{\omega_{\text{sun}}} = \pm \frac{\Pi(N_{\text{driver}})}{\Pi(N_{\text{driven}})} = \left(-\frac{N_s}{N_p}\right)\left(+\frac{N_p}{N_r}\right) = \left(-\frac{N_s}{N_r}\right)$$

Using the concept of relative velocity discussed in Chapter 5, if the arm is allowed to move, then the left side of the equation above must be the ratio of the velocities relative to the arm instead of the absolute velocities; thus

$$\frac{\omega_{\text{ring}} - \omega_{\text{arm}}}{\omega_{\text{sun}} - \omega_{\text{arm}}} = \left(-\frac{N_s}{N_p}\right)\left(+\frac{N_p}{N_r}\right)$$

In general

$$\frac{\omega_{\text{last}} - \omega_{\text{arm}}}{\omega_{\text{first}} - \omega_{\text{arm}}} = \pm \frac{\Pi(N_{\text{driver}})}{\Pi(N_{\text{driven}})} \tag{9.15}$$

Note that this is a vector equation, so you must assign a direction before entering velocities. The right side of the above equation is the product of the ratio of the number of teeth on the driver gears divided by the number of teeth on the driven gears with the appropriate sign inserted. If the two mating gears are externally meshed, then the ratio contains a negative sign since the two gears will rotate in opposite directions. If the two mating gears are internally meshed, then the ratio contains a plus sign since the two gears will rotate the same direction.

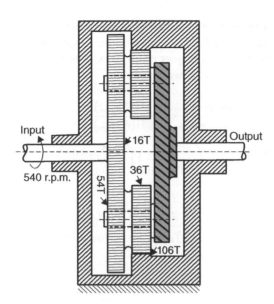

Figure 9.24   Planetary gear train

Example 9.6   Determine the output speed and direction for a planetary gear train using the formula method

Problem: The input shaft rotating at 540 r.p.m. counter clockwise when viewed from the left drives the sun gear which has 16 teeth. A compound gear set having 54 teeth and 36 teeth make up the planet gears. The ring gear has 106 teeth and is fixed relative to the ground. The arm is connected to the output shaft (see Figure 9.24). Determine the output speed using the formula method.

Solution: Note that the sun gear drives the 54-tooth planet gear, the 54-tooth planet gear and the 36-tooth planet gear turn at the same speed, and the 36-tooth planet gear drives the 106-tooth ring gear:

$$(+c.c.w.)\ \frac{\omega_{\text{last}} - \omega_{\text{arm}}}{\omega_{\text{first}} - \omega_{\text{arm}}} = \pm \frac{\Pi(N_{\text{driver}})}{\Pi(N_{\text{driven}})}$$

$$\frac{\omega_{\text{ring}} - \omega_{\text{arm}}}{\omega_{\text{sun}} - \omega_{\text{arm}}} = \left(-\frac{N_{\text{sun16}}}{N_{\text{planet54}}}\right)\left(+\frac{N_{\text{planet36}}}{N_{\text{ring106}}}\right) = \left(-\frac{16}{54}\right)\left(+\frac{36}{106}\right)$$

$$\frac{0 - \omega_{\text{arm}}}{540 - \omega_{\text{arm}}} = -\frac{576}{5724}$$

$$540*576 - 576\omega_{\text{arm}} = 5724\omega_{\text{arm}}$$

$$\omega_{\text{arm}} = \frac{311{,}040}{5724 + 576} = +49.37\,\text{r.p.m.} = \omega_{\text{output}}$$

Answer: $\boxed{\omega_{\text{output}} = 49.4\,\text{r.p.m. c.c.w. from the left}}$.

If the planetary gear system contains two planetary gear trains, then it is referred to as a compound planetary gear system (see Figure 9.25). In this case, the formula method must be applied twice since there are two paths through the gear train. Also, note that there are two separate unknowns, ω_{arm} and ω_{output}; thus, we need two equations:

$$(+c.c.w.)\ \frac{\omega_{last} - \omega_{arm}}{\omega_{first} - \omega_{arm}} = \pm\ \frac{\Pi(N_{driver})}{\Pi(N_{driven})}$$

$$\frac{\omega_{ring120} - \omega_{arm}}{\omega_{sun20} - \omega_{arm}} = \left(-\frac{N_{sun}}{N_{planet1}}\right)\left(+\frac{N_{planet1}}{N_{fixed\_ring}}\right)$$

and

$$\frac{\omega_{ring106} - \omega_{arm}}{\omega_{sun20} - \omega_{arm}} = \left(-\frac{N_{sun}}{N_{planet1}}\right)\left(+\frac{N_{planet2}}{N_{ring\_gear2}}\right)$$

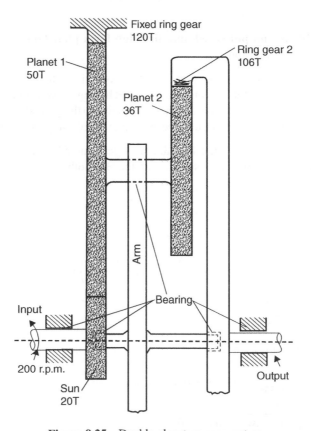

Figure 9.25   Double planetary gear system

9.10.5   Table Method

The table method uses the concept of relative velocity also. It can be applied in four steps. The procedure follows:

Step 1   Hold the arm fixed, and rotate the initially fixed gear one revolution. Calculate the amount each of the other gears rotate, paying close attention to their direction relative to the first gear.

Step 2   Rotate the entire assemble back one revolution. All gears and the arm rotate exactly one revolution back.

Step 3   Add the two rows of values. The originally fixed gear must add to zero.

Step 4   Use the summed relationships to obtain the relative velocities for the remaining gears.

Example 9.7   Determine the output speed and direction for a planetary gear train using the table method

Problem: The input shaft rotating at 540 r.p.m. counter clockwise when viewed from the left drives the sun gear which has 16 teeth. A compound gear set having 54 teeth and 36 teeth make up the planet gears. The ring gear has 106 teeth and is fixed relative to the ground. The arm is connected to the output shaft (see Figure 9.26). Determine the output speed using the table method.

Solution: Note that the sun gear drives the 54-tooth planet gear, the 54-tooth planet gear and the 36-tooth planet gear turn at the same speed, and the 36-tooth planet gear

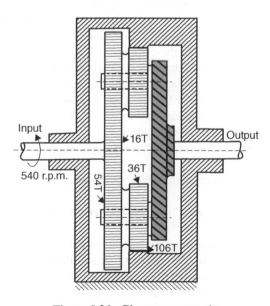

Figure 9.26   Planetary gear train

drives the 106-tooth ring gear that is fixed to ground. Set up the table, fill in the information, sum the two rows, and then calculate the relative velocities between the desired members.

Sun	Planets 1 and 2	Ring	Arm	Condition
(+106/36)(−54/16)	+106/36	+1	0	Arm fixed
−1	−1	−1	−1	All rotate −1
−10.9375	1.9444	0	−1	Sum

The sun must rotate 10.9375 times for each time that the arm rotates in the opposite direction:

$$\frac{\omega_{sun}}{\omega_{arm}} = \frac{-10.9375}{-1}$$

$$\omega_{arm} = 540\,r.p.m.\,c.c.w.$$

$$\omega_{arm} = +49.37 = \omega_{output}$$

Answer: $\boxed{\omega_{output} = 49.4\,r.p.m.\,c.c.w.\,from\,the\,left}$.

Problems

9.1 A standard size pinion with a 3.00 in. pitch circle has an addendum of 0.200 in. How many teeth are on the pinion? What is its circular pitch?

9.2 A 24-tooth pinion with a diametric pitch of 10 teeth/in. and a pressure angle of 20° is to mesh with a similar size gear. Calculate the outside diameter of the pinion, the addendum, the dedendum, the pitch diameter, and the circular pitch.

9.3 A 2-module, 20° pinion with 20 teeth drives a 40-tooth gear. Calculate the pitch radius (mm), base radius (mm), and the circular pitch of each gear.

9.4 A 48-tooth pinion with a pitch radius of 96.0 mm and a pressure angle of 20° is to mesh with a similar size gear. Calculate the outside diameter of the pinion, the addendum, the dedendum, the pitch diameter, and the circular pitch.

9.5 A 35-tooth pinion with a pressure angle of 20° drives an 80-tooth gear. Calculate the contact ratio.

9.6 A 10-tooth pinion with a pressure angle of 25° drives a 20-tooth gear. Calculate the contact ratio.

9.7 Calculate the number of teeth on a 14.5° involute spur gear so that its base circle lines on top of the dedendum circle.

9.8 For a 22.5° pressure angle, calculate the smallest pinion that will mesh with a gear rack without interference. What is the smallest pinion that will mesh with itself?

9.9 An 18-tooth pinion with a diametric pitch of 12 teeth/in. drives a 54-tooth gear. The actual center distance is 3.05 in. Both gears have a pressure angle of 20°. Calculate the operating pressure angle.

9.10   A 2.5-module 18-tooth pinion with a pressure angle of 25° drives a 50-tooth gear. If the center distance at assembly is increased by 1.00 mm, what is the operating pressure angle?

9.11   A 16-tooth pinion with a pressure angle of 20° and a diametric pitch of 8 teeth/in. meshes with a 64-tooth gear. If the center distance at assembly is increased by 0.03 in., what is the operating pressure angle? What are the operating pitch circle diameters for these two meshing gears?

9.12   The center distance between two mating gears is 6.00 in. The gear ratio is 5:1. The diametric pitch is 10 teeth/in. with a pressure angle of 20°. What is the number of teeth on each gear? Is interference present?

9.13   The center distance between two mating gears is 120 mm. The gear ratio is 7:1. The module is 3 mm/tooth with a pressure angle of 25°. What is the number of teeth on each gear? Is interference present?

9.14   The center distance between two mating gears is 5.50 in. The pinion has 15 teeth. The diametric pitch is 5 teeth/in. with a pressure angle of 20°. What is the number of teeth on the larger gear? What is the gear ratio?

9.15   The center distance between two mating gears is 100 mm. The 4-module pinion has 13 teeth. The pressure angle is 25°. What is the number of teeth on the larger gear? What is the gear ratio?

9.16   Analyze the inverted compound gear train shown in Figure 9.27. Assume $N_1 = 20T$, $N_2 = 50\,T$, $N_3 = 16\,T$, and $N_4 = 24\,T$. $P_{d12} = 14$ teeth/in. What is the center distance? What is P_{d34}? If the input speed is 200 r.p.m. clockwise, what are the output speed and direction?

9.17   Analyze the inverted compound gear train shown in Figure 9.27. Assume $N_1 = 15\,T$, $N_2 = 45\,T$, $N_3 = 14\,T$, and $N_4 = 18\,T$. The module for gears 1 and 2 is 2 mm/tooth. What is the center distance? What is the module for gears 3 and 4? If the input speed is 300 r.p.m. counterclockwise (c.c.w.), what are the output speed and direction?

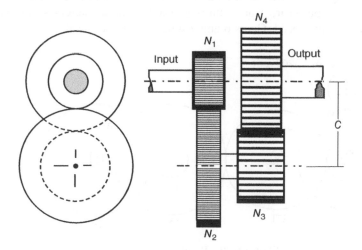

Figure 9.27   Problems 16 and 17

9.18   With an input speed of 900 r.p.m. counterclockwise, determine the output speed and direction for the gear train shown in Figure 9.28. All gears have the same diametric pitch of 10 teeth/in. and the same thickness. What is the center distance between the input shaft and the output shaft if all shafts are in the same plane?

9.19   Can you rearrange the 6 gears found in problem 9.18 (Figure 9.28) so the kinetic energy of the system is minimized? The overall gear ratio must remain the same. Can the overall center distance between the input shaft and the output shaft be reduced by swapping gear locations? All gears have the same diametric pitch and thickness.

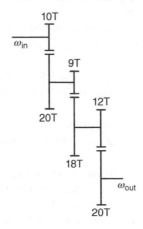

Figure 9.28   Problems 18 and 19

9.20   Determine the angular velocity of the output gear (gear 8) for the gear train shown in Figure 9.29. The input gear, gear 2, turns at 1200 r.p.m. counterclockwise. Gears 3 and 4 turn at the same speed. Gears 6 and 7 turn at the same speed.

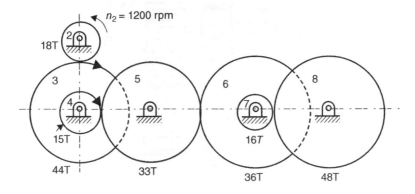

Figure 9.29   Problem 20

9.21   Calculate the equivalent mass moment of inertia for the four different gear trains shown
        in Figure 9.30, and then determine which is the smallest. Assume the mass density of steel is
        0.283 lbm/in.^{3}, and the thickness of all gears is 0.5 in. Assume the input gear is on the left in
        each case and is rotating at 100 rad/s clockwise. (One approach would be to calculate the
        total kinetic energy of each gear system and then set it equal to ½*$I_{equivalent}$ * ω_{in}^2 and solve
        for $I_{equivalent}$.) In all cases, the overall gear ratio is 8:1. Compare the results.

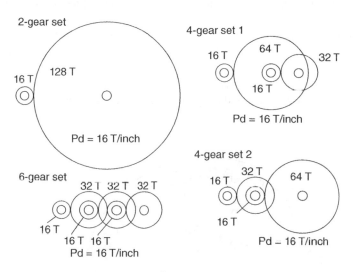

Figure 9.30   Problem 21

9.22   For the planetary gear train shown in Figure 9.31, determine the output speed and
        direction if the input is rotating at 400 r.p.m. clockwise when viewed from the left.
        The second 50-tooth gear is fixed relative to ground. If the diametric pitch for all gears
        is 8 teeth/in., what is the distance between gear centers?

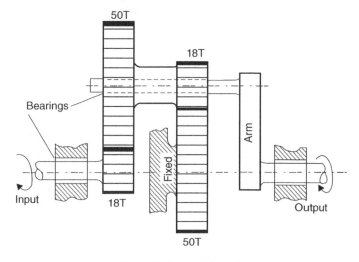

Figure 9.31   Problem 22

9.23   For the planetary gear system shown in Figure 9.32, determine the output speed and direction of the ring gear if the sun gear is held fixed and the arm turns at 15 r.p.m. counterclockwise. The sun gear has 18 teeth and the planet gear has 12 teeth. How many teeth are on the ring gear?

9.24   For the planetary gear system shown in Figure 9.32, determine the output speed and direction of the sun gear if the ring gear is held fixed and the arm turns at 10 r.p.m. clockwise. The sun gear has 18 teeth and the planet gear has 12 teeth. How many teeth are on the ring gear?

9.25   A planetary gear train similar to Figure 9.32 is to be designed using a different number of teeth. If the sun gear has 24 teeth and the planet gear has 13 teeth, how many teeth must the ring gear have?

9.26   A planetary gear train similar to Figure 9.32 is to be designed using a different number of teeth. If the ring gear has 96 teeth and the planet gear has 23 teeth, how many teeth must the sun gear have?

Figure 9.32   Problems 23 and 26

9.27   For the two-motor system shown in Figure 9.33, determine the speed and direction of the drum given that motor #1 drives the 18-tooth 20° sun gear at 500 r.p.m. clockwise, while motor #2 drives a 20-tooth 14.5° spur gear clockwise at 1000 r.p.m. The 20-tooth spur gear drives a 120-tooth spur gear on the outside of the ring gear housing. The internal portion of the ring gear has 100 teeth. The arm (output of the planetary gear system) drives a 16-tooth 25° spur gear that meshes with a 64-tooth spur gear attached to the drum.

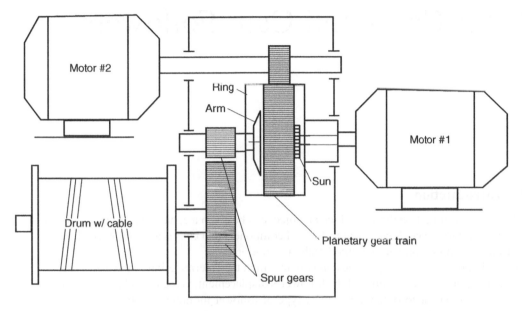

Figure 9.33   Problem 27

9.28   The DC motor shown in Figure 9.34 operates at 1720 r.p.m. An inverted compound gear train is used to reduce the speed of the motor to approximately 320 r.p.m. Gears 1 and 2 have a diametric pitch of 12 teeth/in., while gears 3 and 4 have a diametric pitch of 8 teeth/in. Determine the number of gear teeth on each spur gear for a working design. Note that there is more than one possible answer. (*One possible solution uses 40 teeth on gear 3.*)

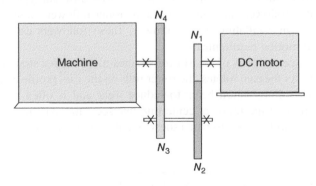

Figure 9.34   Problem 28

10

Planar Cams and Cam Followers

10.1 Introduction

Sometimes, the simplest method by which an object can be given a certain prescribed motion is to use a cam and follower. A cam is a mechanical component that causes another object, known as the follower, to conform to a defined path and motion.

Cam design is ordinarily a syntheses process in which the determination of the cam shape is based on predetermined conditions of the follower's displacement, velocity, and acceleration. It is theoretically possible to obtain almost any type of follower motion by proper design of the cam. However, practical design considerations often necessitate modifications of the desired follower motions.

Although there are a variety of cam types available, we will only discuss planar disk or plate cam motion in this chapter. Practical applications for cams are numerous; one of the better-known applications is the timing system used in gasoline or diesel engines.

A cam mechanism usually consists of the cam (driver), the follower (driven component), and the frame or support for the cam and follower. The follower normally has either translating or rotating motion. Figure 10.1 illustrates this type of cam with four different type followers: an offset translating knife-edge follower, an in-line translating roller follower, an in-line translating flat-face follower, and an oscillating roller follower. These followers can satisfy most common one-dimensional motion requirements.

A translating roller follower consists of an arm constrained to move in a straight line and a roller attached using a pin. As the cam rotates, the roller rolls on the cam profile and causes the follower to translate. This rolling action helps to reduce wear and is often preferred over followers with sliding action. If the follower centerline intersects the axis of rotation of the cam, then the follower is said to be in-line; otherwise, it is considered to be offset. In

Design and Analysis of Mechanisms: A Planar Approach, First Edition. Michael J. Rider.
© 2015 John Wiley & Sons, Ltd. Published 2015 by John Wiley & Sons, Ltd.

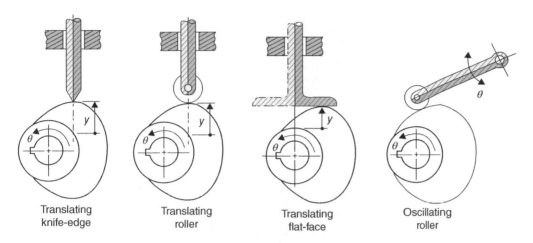

Figure 10.1   Disk cam with different types of followers

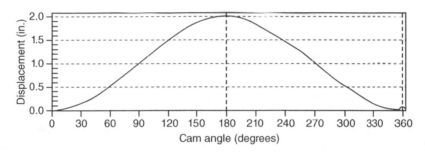

Figure 10.2   Simple harmonic motion of the follower

Figure 10.1, the first two followers are offset and the third follower is in-line. The term offset or in-line does not apply to oscillating followers.

Let us assume that we want the follower to move with simple harmonic motion rising 2 in. during the first 180° of cam rotation and return during the remaining 180° while the cam rotates clockwise (c.w.). We would like this motion to repeat four times a second. The follower motion could be described in terms of time rather than angle of rotation of the cam; however, to design the cam, we need to know the amount of lift at each rotation point of the cam regardless of the cam's speed. The follower displacement diagram provides the first step in cam design (see Figure 10.2).

Before continuing, we need to define some common cam terms. The greatest distance through which the follower moves is known as the total follower stroke. The center of the roller is known as the trace point. The pitch curve is the curve through which the trace point moves if the cam is held stationary and the follower is rotated around the cam. It is necessary to keep the roller follower in contact with the cam surface. This is done by means of a spring force or in some cases the weight of the follower when it moves vertically. However, great care must be exercised in the design of a cam to ensure contact between the follower and the cam at all times.

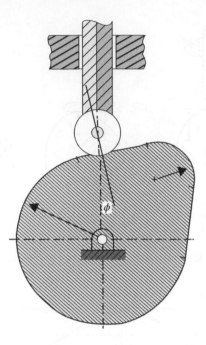

Figure 10.3   Pressure angle, ϕ.

The prime circle is the smallest circle whose center is at the cam's rotation point and is tangent to the pitch curve.

A complication associated with a roller follower is that the point of contact between the roller and the cam does not always lie on the follower's centerline. The force that exists between the follower and the cam acts along a line perpendicular to the tangent between the follower and the cam at the point of contact. The pressure angle of the roller follower is the angle between the line of travel of the follower and the normal drawn to the pitch curve. The pressure angle is known as ϕ (see Figure 10.3). The higher the pressure angle, the higher the bending force applied to the follower. A method to reduce the side force presented to a roller follower is to offset the follower's line of action from the center of rotation of the cam. A flat-face follower has the advantage of 0° pressure angle throughout its motion.

10.2   Follower Displacement Diagrams

Before the cam contour can be determined, it is necessary to select the motion with which the follower will move in accordance with the system requirements. If operation is at a slow speed, the motion may be any one of several common motions, for example, parabolic, harmonic, fifth-order polynomial, 5-4-3 polynomial, or cycloidal.

Parabolic motion has the lowest theoretical acceleration for a given rise and cam speed for the motions listed earlier, and for this reason, it has been used for many cam profiles. Parabolic motion may also be modified to include an interval of constant velocity between the acceleration and deceleration phases. This is often spoken of as a modified constant velocity profile.

However, it is important to note that the jerk or derivative of acceleration with respect to time approaches a very large value at the beginning, midpoint, and end of this interval so parabolic motion should not be used for high-speed cams.

Harmonic motion has the advantage that with a radial roller follower, the maximum pressure angle will be smaller than for parabolic or cycloidal motion with an equal time interval. This will allow the follower to be less rigidly supported, and less power will be needed to operate the cam. However, harmonic motion's acceleration jumps from zero to a very large value almost instantly. Since the derivative of acceleration with respect to time is jerk, harmonic motion will have a very high jerk at its beginning and ending points if the accelerations from the previous and following intervals do not match.

After the follower motion has been selected, it is necessary to determine the follower displacement versus cam angle as shown in Figure 10.4. In this case, the follower dwells for 90°, then rises a specified distance during 90°, then dwells for another 90°, and finally falls back to the starting position during the last 90°.

For cams operating at higher speeds, the selection of the motion of the cam follower must be based not only on the displacement but also on the forces acting on the system as a result of the motion selected. For many years, cam design was concerned only with moving the follower through a given distance in a certain length of time. Speeds were low so the acceleration forces were unimportant. With the trend toward higher machine speeds, it is necessary to consider the dynamic characteristics of the system and to select a cam contour that will minimize the dynamic loading and prevent cam and follower separation.

As an example of the importance of dynamic loading, consider parabolic motion. On the basis of inertia forces, this motion would seem to be very desirable because its acceleration is low. The acceleration increases from zero to a constant value almost instantaneously which results in a high rate of change or jerk. Jerk is an indication of the impact characteristics of the loading. A high jerk value corresponds to a high impact load. Lack of rigidity and backlash in the system tends to increase the effects of impact loading. In parabolic motion where jerk is infinite, this impact occurs twice during the cycle and has the effect of a sharp blow on the system at the beginning and ending points. This may set up undesirable vibrations and cause structural damage.

To avoid infinite jerk, cycloidal or polynomial curve fits should be used for the follower rises and falls. We want acceleration to increase gradually and jerk to be finite. As an example, when a rise follows a dwell, the zero acceleration at the end of the dwell is matched by selecting a curve having zero acceleration at the start of the rise.

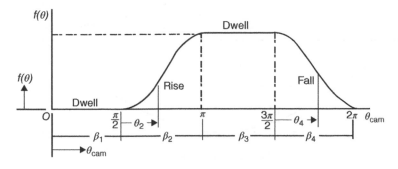

Figure 10.4   Follower displacement diagram

The cycloidal motion provides zero acceleration at both ends of the action. It can be coupled to a dwell at each end. Because the pressure angle is relatively high and the acceleration returns to zero, two cycloidal motions should not follow each other.

The harmonic motion provides the lowest peak acceleration and pressure angle of the three profiles. It is preferred when the acceleration at both the start and the finish can be matched to the end accelerations of the adjacent profiles. Because acceleration at the midpoint is zero, the half harmonic can often be used when a constant velocity rise follows an acceleration. However, a dwell cannot be inserted in this motion because the jerk would be infinite at the transition.

10.3   Harmonic Motion

Harmonic motion works well for cams running at low speeds because of the very large jerk at its beginning and ending points since the acceleration goes from zero to nonzero instantaneously. Do not use this motion before or after a dwell for high-speed cams (see Figure 10.5).

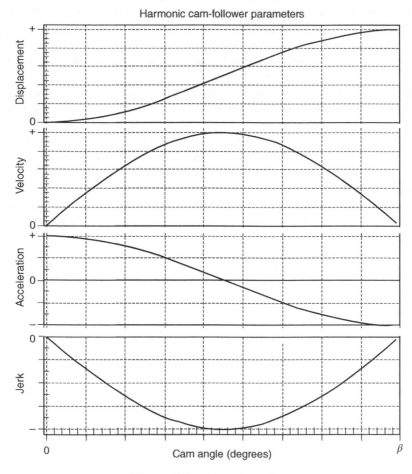

Figure 10.5   Harmonic motion

Harmonic motion can be determined using the following equations. The units for θ and β must be the same units, either degrees or radians. The arguments for the sine and cosine are calculated in radians since calculus is involved. Note that θ ranges from zero to β, where β is assumed to be in degrees:

$$f(\theta) = h_\text{o} + \frac{(\Delta h)}{2}\left(1 - \cos\left(\frac{\pi\theta}{\beta}\right)\right) \tag{10.1}$$

$$f'(\theta) = \frac{\pi(\Delta h)}{2\beta}\sin\left(\frac{\pi\theta}{\beta}\right) \tag{10.2}$$

$$f''(\theta) = \frac{\pi^2(\Delta h)}{2\beta^2}\cos\left(\frac{\pi\theta}{\beta}\right) \tag{10.3}$$

where

$(\Delta h) > 0$ for a rise

$(\Delta h) < 0$ for a fall

Use the following equations to determine the velocity and acceleration of the follower for a given cam speed. Note that the $(180/\pi)$ conversion factor is necessary if β is in degrees:

$$\text{Velocity} = \left(\frac{180 \cdot \omega_\text{cam}}{\pi}\right)f'(\theta) \tag{10.4}$$

$$\text{Acceleration} = \left(\frac{180 \cdot \omega_\text{cam}}{\pi}\right)^2 f''(\theta) \tag{10.5}$$

where

ω_cam is in $\left\{\frac{\text{rad}}{s}\right\}$

10.4   Cycloidal Motion

Cycloidal motion works well for cams running at high speeds because the jerk at its beginning and ending points is finite. Cycloidal motion works well before or after a dwell since its velocity and acceleration start and end at zero (see Figure 10.6).

Cycloidal motion can be determined using the following equations. The units for θ and β must be the same units, either degrees or radians. The arguments for the sine and cosine are calculated in radians since calculus is involved. Note that θ ranges from zero to β, where β is assumed to be in degrees:

where

$(\Delta h) > 0$ for a rise

$(\Delta h) < 0$ for a fall

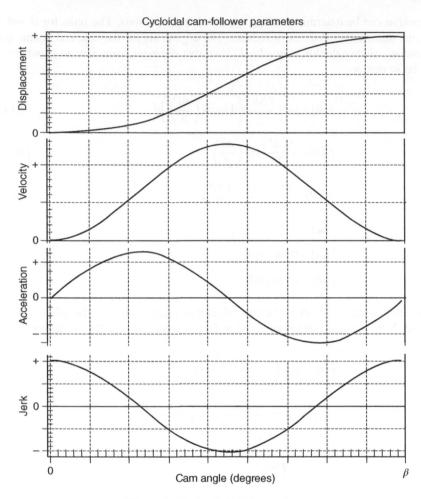

Figure 10.6   Cycloidal rise motion

$$f(\theta) = h_{\mathrm{o}} + (\Delta h)\left(\frac{\theta}{\beta} - \frac{1}{2\pi}\sin\left(\frac{2\pi\theta}{\beta}\right)\right) \tag{10.6}$$

$$f'(\theta) = \frac{(\Delta h)}{\beta}\left(1 - \cos\left(\frac{2\pi\theta}{\beta}\right)\right) \tag{10.7}$$

$$f''(\theta) = \frac{2\pi(\Delta h)}{\beta^2}\left(\sin\left(\frac{2\pi\theta}{\beta}\right)\right) \tag{10.8}$$

Use the following equations to determine the velocity and acceleration of the follower for a given cam speed. Note that the $(180/\pi)$ conversion factor is necessary if β is in degrees:

$$\text{Velocity} = \left(\frac{180 \cdot \omega_{\text{cam}}}{\pi}\right) f'(\theta)$$

$$\text{Acceleration} = \left(\frac{180 \cdot \omega_{\text{cam}}}{\pi}\right)^2 f''(\theta)$$

where

ω_{cam} is in $\left\{\frac{\text{rad}}{\text{s}}\right\}$

10.5   5-4-3 Polynomial Motion

The 5-4-3 polynomial motion works well for cams running at high speeds because the jerk at its beginning and ending points is finite. The 5-4-3 polynomial motion works well before or after a dwell since its velocity and acceleration start and end at zero (see Figure 10.7).

The 5-4-3 polynomial motion can be determined using the following equations. The units for θ and β must be the same units, either degrees or radians. Note that θ ranges from zero to β, where β is assumed to be in degrees:

$$f(\theta) = h_0 + (\Delta h)\left[6\left(\frac{\theta}{\beta}\right)^5 - 15\left(\frac{\theta}{\beta}\right)^4 + 10\left(\frac{\theta}{\beta}\right)^3\right] \tag{10.9}$$

$$f'(\theta) = \frac{(\Delta h)}{\beta}\left[30\left(\frac{\theta}{\beta}\right)^4 - 60\left(\frac{\theta}{\beta}\right)^3 + 30\left(\frac{\theta}{\beta}\right)^2\right] \tag{10.10}$$

$$f''(\theta) = \frac{(\Delta h)}{\beta^2}\left[120\left(\frac{\theta}{\beta}\right)^3 - 180\left(\frac{\theta}{\beta}\right)^2 + 60\left(\frac{\theta}{\beta}\right)\right] \tag{10.11}$$

where

$(\Delta h) > 0$ for a rise

$(\Delta h) < 0$ for a fall

Use the following equations to determine the velocity and acceleration of the follower for a given cam speed. Note that the $(180/\pi)$ conversion is necessary if β is in degrees:

$$\text{Velocity} = \left(\frac{180 \cdot \omega_{\text{cam}}}{\pi}\right) f'(\theta)$$

$$\text{Acceleration} = \left(\frac{180 \cdot \omega_{\text{cam}}}{\pi}\right)^2 f''(\theta)$$

where

ω_{cam} is in $\left\{\frac{\text{rad}}{\text{s}}\right\}$

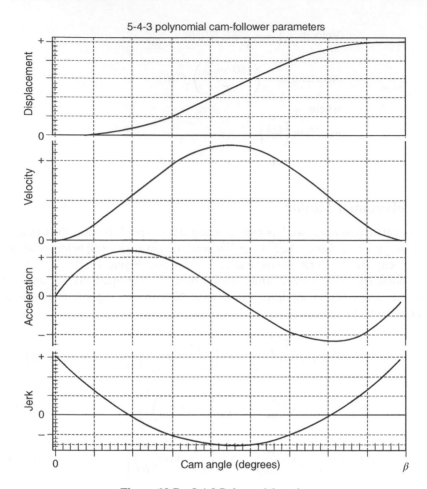

Figure 10.7 5-4-3 Polynomial motion

10.6 Fifth-Order Polynomial Motion

The fifth-order polynomial motion works well for cams running at high speeds because the jerk at its beginning and ending points is finite. The fifth-order polynomial motion works well before or after any other type of motion since its velocity and acceleration can be any value at its start and ending points. If the initial and ending velocities and accelerations are zero, then the fifth-order polynomial motion becomes the 5-4-3 polynomial motion (see Figure 10.8).

The graph earlier represents the follower parameters for a follower that goes from 1 to 5 in 90°. It has to start and end with a velocity of 2. It starts with an acceleration of 1 and ends with an acceleration of zero. The jerk needs to be continuous and finite:

$$f(\theta) = \left[11.5 \left(\frac{\theta}{\beta} \right)^5 - 28.5 \left(\frac{\theta}{\beta} \right)^4 + 18.5 \left(\frac{\theta}{\beta} \right)^3 + 0.50 \left(\frac{\theta}{\beta} \right)^2 + 2.0 \left(\frac{\theta}{\beta} \right) + 1.0 \right] \qquad (10.12)$$

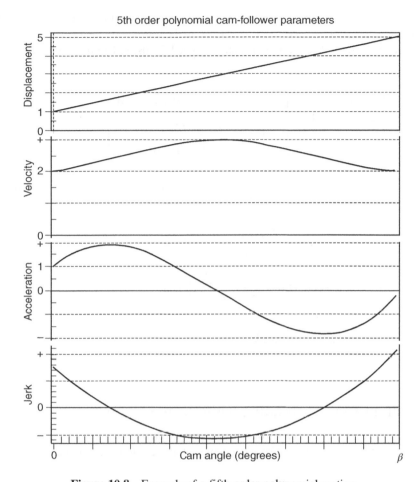

Figure 10.8   Example of a fifth-order polynomial motion

The fifth-order polynomial motion can be determined using the following equations. The units for θ and β must be the same units, either degrees or radians. Note that θ ranges from zero to β, where β is assumed to be in degrees:

$$f(\theta) = \left[c_5 \left(\frac{\theta}{\beta} \right)^5 + c_4 \left(\frac{\theta}{\beta} \right)^4 + c_3 \left(\frac{\theta}{\beta} \right)^3 + \frac{f''(0)}{2} \left(\frac{\theta}{\beta} \right)^2 + f'(0) \left(\frac{\theta}{\beta} \right) + f(0) \right] \qquad (10.13)$$

$$f'(\theta) = \frac{1}{\beta} \left[5c_5 \left(\frac{\theta}{\beta} \right)^4 + 4c_4 \left(\frac{\theta}{\beta} \right)^3 + 3c_3 \left(\frac{\theta}{\beta} \right)^2 + f''(0) \left(\frac{\theta}{\beta} \right) + f'(0) \right] \qquad (10.14)$$

$$f''(\theta) = \frac{1}{\beta^2} \left[20c_5 \left(\frac{\theta}{\beta} \right)^3 + 12c_4 \left(\frac{\theta}{\beta} \right)^2 + 6c_3 \left(\frac{\theta}{\beta} \right) + f''(0) \right] \qquad (10.15)$$

To solve for c_5, c_4, and c_3, we need to apply three additional conditions, such as the end conditions of the interval, $f(\beta)$, $f'(\beta)$, and $f''(\beta)$. This leads to the following linear system of three equations and three unknowns. The unknowns can be solved for using Cramer's Rule or your calculator:

$$\begin{bmatrix} 1 & 1 & 1 \\ 5 & 4 & 3 \\ 20 & 12 & 6 \end{bmatrix} \begin{Bmatrix} c_5 \\ c_4 \\ c_3 \end{Bmatrix} = \begin{Bmatrix} f(\beta) - f(0) - f'(0) - \frac{1}{2} f''(0) \\ f'(\beta) - f'(0) - f''(0) \\ f''(\beta) - f''(0) \end{Bmatrix}$$

Use the following equations to determine the velocity and acceleration of the follower for a given cam speed. Note that the $(180/\pi)$ conversion is necessary if β is in degrees:

$$\text{Velocity} = \left(\frac{180 \cdot \omega_{\text{cam}}}{\pi} \right) f'(\theta)$$

$$\text{Acceleration} = \left(\frac{180 \cdot \omega_{\text{cam}}}{\pi} \right)^2 f''(\theta)$$

where

$$\omega_{\text{cam}} \text{ is in } \left\{ \frac{\text{rad}}{\text{s}} \right\}$$

10.7 Cam with In-Line Translating Knife-Edge Follower

The analytical determination of the pitch surface for a cam with an in-line knife-edge follower is shown in Figure 10.9. "R" is the distance from the center of the cam to the tip of the knife-edge follower:

$$R = R_b + f(\theta)$$

R_b = base radius of cam (constant)

$f(\theta)$ = radial motion of the follower as a function of cam angle

The pressure angle, ϕ, is the angle between a line drawn perpendicular to the follower motion and the slope of the cam surface at the point of contact. The pressure angle can be determined by noting that as the cam rotates through an arc length of $Rd\theta$, the follower moves a distance of dR:

$$\tan\phi = \frac{dR}{Rd\theta}$$

$$\phi = \tan^{-1}\left(\frac{1}{R} \cdot \frac{dR}{d\theta} \right) = \tan^{-1}\left(\frac{f'(\theta)}{R} \right) = \tan^{-1}\left(\frac{f'(\theta)}{R_b + f(\theta)} \right)$$

We can calculate the point of the cam surface where the knife-edge follower touches and thus define the shape of the cam itself. Since the follower is directly above the center of the cam, we need to add 90 degrees to the follower's angle to obtain the correct position on the cam surface:

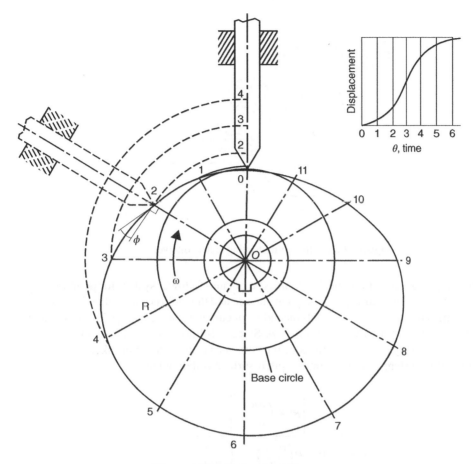

Figure 10.9   Knife-edge follower

$$x_{\text{cam}} = R\cos(\theta + 90°) = -(R_b + f(\theta))\sin\theta \tag{10.16}$$

$$y_{\text{cam}} = R\sin(\theta + 90°) = (R_b + f(\theta))\cos\theta \tag{10.17}$$

10.8   Cam with In-Line Translating Roller Follower

The analytical determination of the pitch surface for a cam with an in-line roller follower is shown in Figure 10.10. R is the distance from the center of rotation of the cam to the center of the roller follower:

$$R = R_b + R_r + f(\theta)$$

R_b = base radius of cam (constant)
R_r = roller radius of follower (constant)
$f(\theta)$ = radial motion of the follower as a function of the cam angle

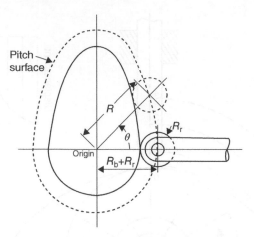

Figure 10.10   In-line translating roller follower

The following is a method for checking this type of cam–follower system for pointing. This involves the radius of curvature, ρ, of the pitch surface and the radius of the roller follower, R_r. The cam becomes pointed when R_r equals the radius of curvature of the cam, ρ. To avoid undercutting the cam, R_r must always be less than ρ. Since it is impossible to undercut a concave portion of the cam, only convex portions of the cam surface need to be checked.

The radius of curvature at a point on the pitch surface is defined as

$$\rho = \frac{\left[R^2 + \left(\dfrac{dR}{d\theta}\right)^2\right]^{1.5}}{R^2 - R\left(\dfrac{d^2R}{d\theta^2}\right) + 2\left(\dfrac{dR}{d\theta}\right)^2}$$

where R is a function of theta and its first two derivatives are continuous. Note that

$$R = R_b + R_r + f(\theta)$$

$$\frac{dR}{d\theta} = f'(\theta)$$

$$\frac{d^2R}{d\theta^2} = f''(\theta)$$

thus

$$\rho = \frac{\left\{R^2 + [f(\theta)]^2\right\}^{1.5}}{R^2 - R \cdot f''(\theta) + 2[f'(\theta)]^2} > R_r$$

To avoid undercutting the cam, R_r must always be less than ρ. Since $f(\theta)$ is always greater than or equal to zero and R is greater than R_r, there is no reason to ever check for pointing during a dwell.

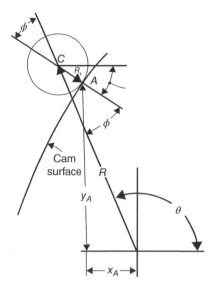

Figure 10.11   Pressure angle for in-line roller follower

As mentioned previously, the pressure angle is an important consideration when designing a roller-follower cam. It is necessary to keep the maximum pressure angle as small as possible, and its maximum should be less than 30°. Although it is possible to make a layout of the cam and measure the maximum pressure angle, analytical methods are preferred.

For a cam with a radial roller follower, the pressure angle ϕ is shown in Figure 10.11. As the cam rotates through an arc length of $Rd\theta$, the follower moves a distance of dR; thus,

$$\tan\phi = \frac{dR}{Rd\theta}$$

If we take the arctangent, we can solve for ϕ. Note that if β was in degrees, then $f'(\theta)$ needs to multiplied by $180/\pi$:

$$\phi = \tan^{-1}\left(\frac{dR}{Rd\theta}\right) = \tan^{-1}\left(\frac{1}{R}\frac{dR}{d\theta}\right) = \tan^{-1}\left(\frac{f'(\theta)}{R_b + R_r + f(\theta)}\right)$$

Once this is done, we can calculate the location of the tangent point, $A = (x_A, y_A)$:

$$\bar{A} = \bar{R} + \bar{R}_r$$

$$x_A = R\cos\theta + R_r\cos(\theta + \phi - \pi)$$

$$y_A = R\sin\theta + R_r\sin(\theta + \phi - \pi)$$

However, the equation can be simplified by using the sine and cosine equivalent for the difference of two angles:

$$\cos(\beta - \pi) = \cos\beta\cos\pi + \sin\beta\sin\pi = -\cos\beta$$

$$\sin(\beta - \pi) = \sin\beta\cos\pi - \cos\beta\sin\pi = -\sin\beta$$

Then

$$x_A = R\cos\theta - R_r \cos(\theta + \phi)$$

$$y_A = R\sin\theta - R_r \sin(\theta + \phi)$$

or

$$x_A = (R_b + R_r + f(\theta))\cos\theta - R_r \cos(\theta + \phi) \tag{10.18}$$

$$y_A = (R_b + R_r + f(\theta))\sin\theta - R_r \sin(\theta + \phi) \tag{10.19}$$

Example 10.1   Determine if the roller radius for an in-line translating roller follower is acceptable and verify that the pressure angle does not exceed 30°. Design the cam

Problem: Design a plate cam and in-line, translating roller follower that rises 1.50 in. during the first 120°, dwells for 30°, falls 1.50 in. during the next 180°, and then dwells for the remainder of the cycle. The cam rotates at 20 rad/s clockwise. The roller diameter is 1.00 in.

Procedure:

1. Determine the type of motion for the rise and fall.
2. Create the follower displacement diagram.
3. Determine the base circle's size.
4. Apply the necessary equations to calculate the profile of the cam surface.
5. Check at all points that the pressure angle stays between −30° and +30°.
6. Calculate points on the cam profile.
7. Check at all points that the roller radius is less than the radius of curvature.
8. Plot the cam profile.

Solution: For the rise, let's use cycloidal motion, and for the fall, let's use 5-4-3 polynomial motion:

$$(0 < \theta < 120°, \beta = 120°, \Delta h = 1.50 \text{ in.})$$

$$f(\theta) = 0 + (1.50)\left(\frac{\theta}{120°} - \frac{1}{2\pi}\sin\left(\frac{2\pi\theta}{120°}\right)\right)$$

$$f'(\theta) = \frac{(1.50)}{120°}\left(1 - \cos\left(\frac{2\pi\theta}{120°}\right)\right)$$

$$f''(\theta) = \frac{2\pi(1.50)}{(120°)^2}\left(\sin\left(\frac{2\pi\theta}{120°}\right)\right)$$

$$(120° < \theta < 150°, \beta = 30°, \text{ dwell at } 1.50 \text{ in.})$$

$$f(\theta - 120°) = 1.50$$

$$f'(\theta - 120°) = 0$$

$$f''(\theta - 120°) = 0$$

$(150° < \theta < 330°, \beta = 180°, \Delta h = -1.50 \text{ in.})$

$$f(\theta-150°) = 1.50 + (-1.50)\left[6\left(\frac{\theta-150°}{180°}\right)^5 - 15\left(\frac{\theta-150°}{180°}\right)^4 + 10\left(\frac{\theta-150°}{180°}\right)^3\right]$$

$$f'(\theta-150°) = \frac{(-1.50)}{180°}\left[30\left(\frac{\theta-150°}{180°}\right)^4 - 60\left(\frac{\theta-150°}{180°}\right)^3 + 30\left(\frac{\theta-150°}{180°}\right)^2\right]$$

$$f''(\theta-150°) = \frac{(-1.50)}{(180°)^2}\left[120\left(\frac{\theta-150°}{180°}\right)^3 - 180\left(\frac{\theta-150°}{180°}\right)^2 + 60\left(\frac{\theta-150°}{180°}\right)\right]$$

$(330° < \theta < 360°, \beta = 30°, \text{dwell at } 0.00 \text{ in.})$

$f(\theta-330°) = 0.00$

$f'(\theta-330°) = 0.00$

$f''(\theta-330°) = 0.00$

The follower displacement diagram is shown in Figure 10.12. The vertical lines indicate the transitions between two different motion types.

Assume the base circle is 4.00 in. There are many choices for this design variable.

Now calculate the cam profile using the following equations. Note that since β was in degrees, $f'(\theta)$ and $f''(\theta)$ need additional conversion factors. This was done in MATLAB and is not shown here. In the same program, calculate and display the pressure angle at each point along the cam profile. The pressure angle ranges from $-24.3°$ at $\theta = 54°$ to $15.8°$ at $\theta = 250°$. This is acceptable:

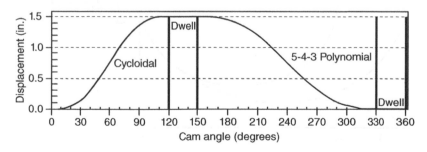

Figure 10.12   Follower displacement diagram

$$\phi = \tan^{-1}\left(\frac{\left(\frac{180}{\pi}\right)f'(\theta)}{2.00+0.50+f(\theta)}\right)$$

$$x_A = (2.00+0.50+f(\theta))\cos\theta-(0.50)\cos(\theta+\phi)$$

$$y_A = (2.00+0.50+f(\theta))\sin\theta-(0.50)\sin(\theta+\phi)$$

$$\rho = \frac{\left\{(2.00+0.50+f(\theta))^2+[f(\theta)]^2\right\}^{1.5}}{(2.00+0.50+f(\theta))^2-(2.00+0.50+f(\theta))\cdot\left(\left(\frac{180}{\pi}\right)^2 f''(\theta)\right)+2\left[\left(\frac{180}{\pi}\right)f'(\theta)\right]^2}$$

In the same program, calculate and display the radius of curvature at each point along the cam profile. The minimum radius of curvature turns out to be 2.45 in. at $\theta = 68°$. This value is greater than the roller radius so this is acceptable.

Plot the cam profile. You can include the base circle and the followers as shown in Figure 10.13.

Answer: $\boxed{\text{Figure 10.13 is an acceptable cam design}}$.

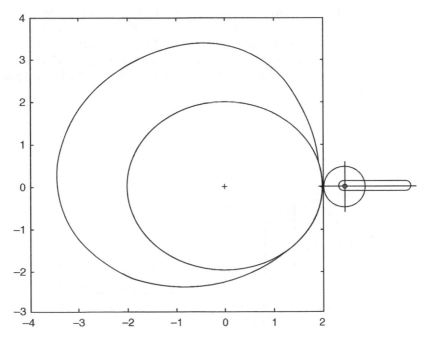

Figure 10.13 Cam profile with roller follower

10.9   Cam with Offset Translating Roller Follower

The roller follower can also be placed away from the center of rotation of the cam. In this case, it is referred to as an offset roller follower (see Figure 10.14). This feature adds an additional design parameter, that is, the amount of follower offset, e. The pressure angle ϕ is again limited to a maximum of 30°.

With the follower moving vertically, it can be shown that the instant center for the cam and the follower is located at the intersection of a horizontal line through the cam rotation point and the line normal to the cam surface at the point of contact. The velocity vector for the follower and this intersection point is exactly the same in magnitude and direction. The distance from the cam rotation point to the instant center is

$$\text{Distance} = e + (d + f(0))\tan\phi$$

where

$$d = \sqrt{(R_b + R_r)^2 - e^2}$$

Then the velocity of the offset follower is

$$V_f = (\text{distance}) \cdot \omega_{\text{cam}} = [e + (d + f(\theta))\tan\phi]\omega_{\text{cam}}$$

The development of the pressure angle equation for offset roller followers is not shown here. However, with the offset, e, defined as positive to the left or upward as shown in Figure 10.14, the pressure angle equation is

$$\phi = \tan^{-1}\left(\frac{f'(\theta) - e}{f(\theta) + \sqrt{(R_b + R_r)^2 - e^2}}\right) \tag{10.20}$$

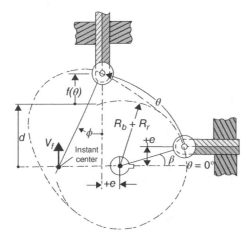

Figure 10.14   Offset roller follower

It is left for the reader to develop the necessary equations to define the cam profile equations. The procedure is similar to the procedure used for the in-line roller follower.

10.10   Cam with Translating Flat-Face Follower

The treatment of the flat-face follower allows the actual cam outlined to be determined analytically. This provides three valuable characteristics for the cam. These are the parametric equation for the cam contour, the minimum radius of the cam to avoid cusps, and the location of the contact point which gives the minimum length of the follower face. Since the value for d can be plus or minus, the minimum length of the flat-face follower should be at least 10% more than $2d$.

Figure 10.15 shows a cam with a radial flat-face follower. The cam rotates at a constant angular velocity clockwise. The contact point between the cam and the follower is at (x, y), which is a distance d from the radial centerline of the follower. The displacement of the follower from the origin is given by the following equation:

$$R = R_b + f(\theta)$$

R_b = radius of the base circle and constant
$f(\theta)$ = the desired motion of the follower as a function of the angular displacement of the cam
From the figure, the (x, y) position of the contact point can be determined in terms of R and d:

$$x = R\cos\theta - d\sin\theta$$
$$y = R\sin\theta + d\cos\theta$$

Rearranging these equations and solving for R and d in terms of x and y lead to

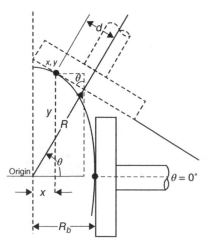

Figure 10.15   Translating flat-face follower

$$R = x\cos\theta + y\sin\theta$$

$$d = -x\sin\theta + y\cos\theta$$

It is interesting to note that the derivative of R with respect to θ is also equal to d so that the x, y equations can be rewritten as

$$\frac{dR}{d\theta} = \frac{df(\theta)}{d\theta} = f'(\theta) = -x\sin\theta + y\cos\theta = d$$

then

$$x = (R_b + f(\theta))\cos\theta - f'(\theta)\sin\theta \qquad (10.21)$$

$$y = (R_b + f(\theta))\sin\theta + f'(\theta)\cos\theta \qquad (10.22)$$

The radius of curvature for a flat-face translating follower cam system is

$$\rho = R_b + f(\theta) + f''(\theta)$$

The minimum radius, R_b, to avoid a cusp or point on the cam surface can be determined analytically. Cusps occur when both $dx/d\theta$ and $dy/d\theta = 0$. If we differentiate the two equations above, we get

$$\frac{dx}{d\theta} = f'(\theta)\cos\theta - (R_b + f(\theta))\sin\theta - f''(\theta)\sin\theta - f'(\theta)\cos\theta = 0$$

$$\frac{dy}{d\theta} = f'(\theta)\sin\theta + (R_b + f(\theta))\cos\theta + f''(\theta)\cos\theta - f'(\theta)\sin\theta = 0$$

thus

$$[R_b + f(\theta) + f''(\theta)]\sin\theta = 0$$

$$[R_b + f(\theta) + f''(\theta)]\cos\theta = 0$$

The first and fourth terms in both equations add to zero and the second and third terms can be combined. Thus, the only way that for both equations to be zero is for

$$[R_b + f(\theta) + f''(\theta)] = 0$$

To avoid cusps,

$$R_b > f(\theta) + f''(\theta)$$

Example 10.2   Determine the minimum face width for a flat-face follower and verify that the cam profile does not contain a cusp. Design the cam

Problem: Design a plate cam with an in-line, translating flat-face follower that rises 1.75 in. during the first 150°, dwells for 60°, falls 1.75 in. during the next 120°, and then dwells for the remainder of the cycle. The cam rotates at 2 rad/s clockwise.

Procedure:

1. Determine the type of motion for the rise and fall.
2. Create the follower displacement diagram.
3. Determine the base circle's size.
4. Apply the necessary equations to calculate the profile of the cam surface.
5. Calculate points on the cam profile.
6. Check at all points to determine the minimum face width for the flat-face follower.
7. Check for cusps.
8. Plot the cam profile.

Solution: For the rise and fall, let's use harmonic motion:

$$(0 < \theta < 150°, \beta = 150°, \Delta h = 1.75 \text{ in.})$$

$$f(\theta) = \frac{(1.75)}{2}\left(1 - \cos\left(\frac{\pi\theta}{150°}\right)\right)$$

$$f'(\theta) = \frac{\pi(1.75)}{2(150°)}\sin\left(\frac{\pi\theta}{150°}\right)$$

$$f''(\theta) = \frac{\pi^2(1.75)}{2(150°)^2}\cos\left(\frac{\pi\theta}{150°}\right)$$

$$(150° < \theta < 210°, \beta = 60°, \text{ dwell at } 1.75 \text{ in.})$$

$$f(\theta - 150°) = 1.75$$

$$f'(\theta - 150°) = 0$$

$$f''(\theta - 150°) = 0$$

$$(210° < \theta < 330°, \beta = 120°, \Delta h = -1.75 \text{ in.})$$

$$f(\theta - 210°) = 1.75 + \frac{(-1.75)}{2}\left(1 - \cos\left(\frac{\pi(\theta - 210°)}{210°}\right)\right)$$

$$f'(\theta) = \frac{\pi(1.75)}{2(210°)}\sin\left(\frac{\pi(\theta - 210°)}{210°}\right)$$

$$f''(\theta) = \frac{\pi^2(1.75)}{2(210°)^2}\cos\left(\frac{\pi(\theta - 210°)}{210°}\right)$$

$$(330° < \theta < 360°, \beta = 30°, \text{ dwell at } 0.00 \text{ in.})$$

$$f(\theta - 330°) = 0.00$$

$$f'(\theta - 330°) = 0.00$$

$$f''(\theta - 330°) = 0.00$$

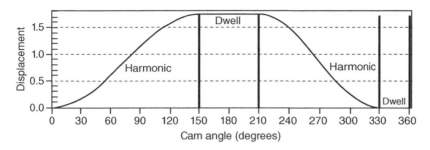

Figure 10.16   Follower displacement diagram

The follower displacement diagram is shown in Figure 10.16. The vertical lines indicate the transitions between two different motion types.

Assume the base circle is 4.00 in. There are many choices for this design variable.

Now calculate the cam profile using the following equations. Note that since β was in degrees, $f'(\theta)$ needs an additional conversion factor. This was done in MATLAB and is not shown here. In the same program, calculate and display the follower width at each point along the cam profile:

$$x = (2.00 + f(\theta))\cos\theta - \left(\frac{180}{\pi}\right)f'(\theta)\sin\theta$$

$$y = (2.00 + f(\theta))\sin\theta + \left(\frac{180}{\pi}\right)f'(\theta)\cos\theta$$

$$R = x\cos\theta + y\sin\theta$$

$$d = -x\sin\theta + y\cos\theta$$

In the same program, calculate and display the minimum follower width at each point along the cam profile. The minimum width turns out to be 2.45 in. at $\theta = 68°$.

In the same program, calculate the radius of curvature for the cam. Note that since β was in degrees, $f''(\theta)$ needs an additional conversion factor:

$$\rho = R_b + f(\theta) + \left(\frac{180}{\pi}\right)^2 f''(\theta)$$

Also, check the following inequality. To avoid cusps,

$$R_b > f(\theta) + \left(\frac{180}{\pi}\right)^2 \cdot f''(\theta)$$

Plot the cam profile. You can include the base circle and the followers as shown in Figure 10.17.

Answer: Figure 10.17 is an acceptable cam design .

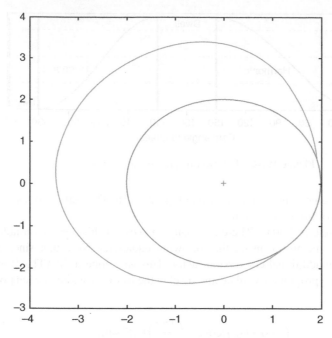

Figure 10.17 Cam profile with flat-face follower

Problems

10.1 Design a double-dwell cam to move a follower from 0 to 2.0 in. in 75°, dwell for 120°, fall 2.0 in. in 75°, and dwell for the remainder. The total cycle must take 3.0 s. Use cycloidal motion for the rise and fall to minimize accelerations. Plot the SVAJ diagrams.

10.2 Design a double-dwell cam to move a follower from 0 to 50 mm in 90°, dwell for 90°, fall 50 mm in 90°, and dwell for the remainder. The total cycle must take 6.0 s. Use harmonic motion for the rise and fall. Plot the SVAJ diagrams.

10.3 Design a double-dwell cam to move a follower from 0 to 3.0 in. in 90°, dwell for 120°, fall 3.0 in. in 120°, and dwell for the remainder. The total cycle must take 5.0 s. Use 5-4-3 polynomial motion for the rise and fall. Plot the SVAJ diagrams.

10.4 Design a double-dwell cam to move a follower from 0 to 20 mm in 30° using harmonic motion, move the follower from 20 to 60 mm in 75° using the fifth-order polynomial, dwell for 75°, fall 60 mm in 120° using cycloidal motion, and dwell for the remainder. The total cycle must take 2.0 s. Be sure to match the velocity and acceleration at each end of the polynomial rise. Plot the SVAJ diagrams.

10.5 Use the cam–follower profile from problem 10.1 to create the cam profile for a vertical translating, knife-edge follower. Let the base circle be equal to 5.0 inches.

10.6 Use the cam–follower profile from problem 10.2 to create the cam profile for a vertical translating, in-line, knife-edge follower. Let the base circle be equal to 60 mm.

10.7   Use the cam–follower profile from problem 10.3 to create the cam profile for a vertical translating, in-line, knife-edge follower. Let the base circle be equal to 4.5 inches.

10.8   Use the cam–follower profile from problem 10.4 to create the cam profile for a vertical translating, in-line, knife-edge follower. Let the base circle be equal to 120 mm.

10.9   Use the cam–follower profile from problem 10.1 to create the cam profile for a translating, in-line, roller follower.

10.10   Use the cam–follower profile from problem 10.2 to create the cam profile for a translating, in-line, roller follower.

10.11   Use the cam–follower profile from problem 10.3 to create the cam profile for a translating, in-line, roller follower.

10.12   Use the cam–follower profile from problem 10.4 to create the cam profile for a translating, in-line, roller follower.

10.13   Use the cam–follower profile from problem 10.1 to create the cam profile for a translating, in-line, flat-face follower.

10.14   Use the cam–follower profile from problem 10.2 to create the cam profile for a translating, in-line, flat-face follower.

10.15   Use the cam–follower profile from problem 10.3 to create the cam profile for a translating, in-line, flat-face follower.

10.16   Use the cam–follower profile from problem 10.4 to create the cam profile for a translating, in line, flat face follower.

Appendix A

Engineering Equation Solver

EES (pronounced "ease") is an acronym for Engineering Equation Solver. The basic function provided by EES is the solution to a set of algebraic equations. EES can solve equations with complex variables and differential equations, do optimization, provide linear and nonlinear regression, generate publication-quality plots, simplify uncertainty analyses, and provide animations. EES has been developed to run under the 32- and 64-bit Microsoft Windows operating systems, that is, Windows 95/98/2000/XP and Windows 7 and 8. It can be run in Linux and on the Macintosh using emulation programs like VMware©.

EES is particularly useful for design problems in which the effects of one or more parameters need to be determined. The program provides this capability with its Parametric Table, which is similar to a spreadsheet. The user identifies the variables that are independent by entering their values in the table cells. EES will calculate the values of the dependent variables in the table. The relationship of the variables in the table can be displayed in graphs. With EES, it is no more difficult to do design problems than it is to solve a problem for a fixed set of independent variables.

EES offers the advantages of a simple set of intuitive commands that a novice can quickly learn to use for solving any algebraic problems. EES can be used to solve many engineering applications; it is ideally suited for instruction in mechanical engineering courses and for the practicing engineer faced with the need to solve practical problems.

Detailed help is available at any point in EES. Pressing the <F1> key will bring up a Help window relating to the frontmost window. Clicking the Contents button will present the Help index. Clicking on an underlined word will provide help relating to that subject.

Starting EES ᴱᴱˢ

The default installation program will create a directory named C:\EES32 in which the EES files are placed. The EES program icon shown above will identify both the program and EES files.

Design and Analysis of Mechanisms: A Planar Approach, First Edition. Michael J. Rider.
© 2015 John Wiley & Sons, Ltd. Published 2015 by John Wiley & Sons, Ltd.

Figure A.1   EES toolbar

Double-clicking the left mouse button on the EES program or file icon will start the program. If you double-clicked on an EES file, that file will be automatically loaded.

Note that a toolbar (Figure A.1) is provided below the menu bar. The toolbar contains small buttons that provide rapid access to many of the most frequently used EES menu commands. If you move the cursor over a button and wait for a few seconds, a few words will appear to explain the function of that button. The toolbar can be hidden, if you wish, with a control in the Preferences dialog (Options menu).

The **System** menu represented by the EES icon appears above the file menu. The System menu is not part of EES, but rather a feature of the Windows operating system. It holds commands that allow window moving, resizing, and switching to other applications.

The **File** menu provides commands for loading, merging and saving work files and libraries, and printing.

The **Edit** menu provides the editing commands to cut, copy, and paste information.

The **Search** menu provides Find and Replace commands for use in the Equations window.

The **Options** menu provides commands for setting the guess values and bounds of variables, the unit system, default information, and program preferences. A command is also provided for displaying information on built-in and user-supplied functions.

The **Calculate** menu contains the commands to check, format, and solve the equation set.

The **Tables** menu contains commands to set up and alter the contents of the Parametric and Lookup Tables and to do linear regression on the data in these tables. The Parametric Table, similar to a spreadsheet, allows the equation set to be solved repeatedly while varying the values of one or more variables. The Lookup Table holds user-supplied data that can be interpolated and used in the solution of the equation set.

The **Plot** menu provides commands to modify an existing plot or prepare a new plot of data in the Parametric, Lookup, or Array Tables. Curve-fitting capability is also provided. The Windows menu provides a convenient method of bringing any of the EES windows to the front or to organize the windows.

The **Help** menu provides commands for accessing the online help documentation (see Figure A.2).

The equations can now be entered into the Equations window. Text is entered in the same manner as for any word processor. Formatting rules are as follows:

1. Upper- and lowercase letters are not distinguished. EES will (optionally) change the case of all variables to match the manner in which they first appear.
2. Blank lines and spaces may be entered as desired since they are ignored.
3. Comments must be enclosed within braces { } or within quotation marks " ". Comments may span as many lines as needed. Comments within braces may be nested in which case only the outermost set of { } are recognized. *Comments within quotes will be displayed in the Formatted Equations window.*
4. Variable names must start with a letter and consist of any keyboard characters except ( ), ', |, *, /, +, −, ^, { }, :, ", or ;. Array variables are identified with square braces around the array

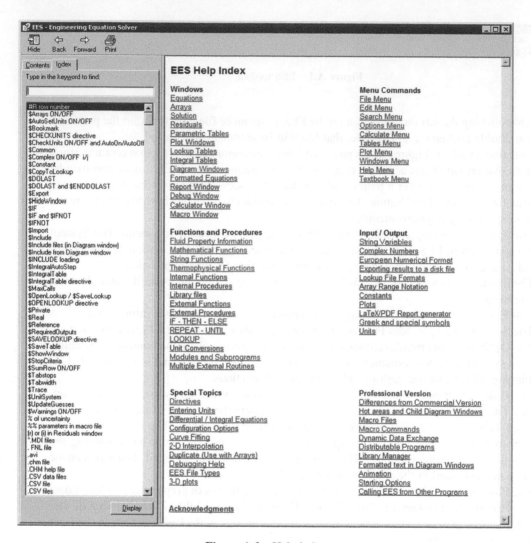

Figure A.2   Help index

index or indices, for example, X[5,3]. String variables are identified with a $ as the last character in the variable name. The maximum length of a variable name is 30 characters.

5. Multiple equations may be entered on one line if they are separated by a semicolon (;). The maximum line length is 255 characters.

6. The caret symbol ^ or ** is used to indicate raising to a power.

7. The order in which the equations are entered does not matter.

8. The position of the knowns and unknowns in the equation does not matter.

9. Units for constants can be entered in braces directly following the comment, for example, g = 9.81 [m/s^ 2].

10. Underscore characters can be used to produce subscripts on the formatted output. Greek symbol names will be replaced by the Greek characters.

11. A vertical bar character in a variable name signifies the start of a superscript.

Unit consistency is as important as entering the correct equations. EES can check the unit consistency of the equations provided that the units of each variable are known. The units of a constant can be set by following the value with its units in square brackets [inch]. The Variable Info menu item provides an easy way to enter the units of a variable in the Equations window.

Before solving nonlinear equations, it is a good idea to set the guess values and (possibly) the lower and upper bounds for the unknown variables before attempting to solve the equations. This is done with the Variable Information command in the Options menu.

The basic capability provided by EES is the solution of a set of linear or nonlinear algebraic equations. To demonstrate this capability, start EES and enter these example problems.

Example A.1   Find (x,y) of vector

Given: A = 4.50 inches at 12 degrees

Find: x and y position of this vector

Solution: Ax = A*cos(12°), and Ay = A*sin(12°)

Start EES, and pick Continue (see Figure A.3).
>Options> Unit System, select English, and pick **OK** (note angles in degrees; see Figure A.4).
In the Equations window, enter the known variables with units in square brackets and the equations to be solved in any order. Comments are added by beginning a line with a double slash (//) and brackets { } or enclosing the text in double quotes. The second option will display the comments with the formatted equations when printed. Greek letters are entered by typing out their name such as beta, gamma, theta, etc. Subscripts are added to a variable name by following the name with an underscore and the subscript value such as beta_1 (β_1).

Example: Enter the equations in the Equations window (see Figure A.5).

>Calculate>Solve.

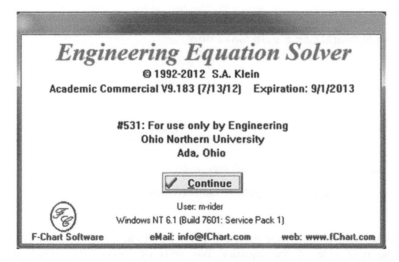

Figure A.3   EES start-up screen

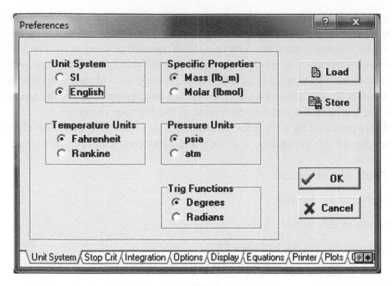

Figure A.4   Preference window

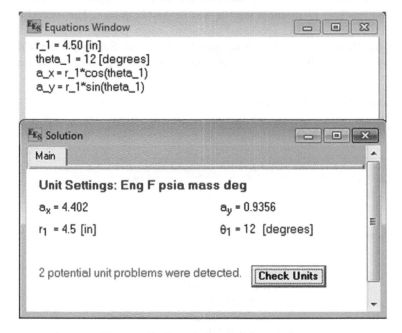

Figure A.5   Equations and Solve window

Note that two potential problems were detected. The variables a_x and a_y were not assigned any units, but the right side of the equation has units of inches. Right-click in the Equations window, and then select Variable Info (see Figure A.6). In the Units column, add

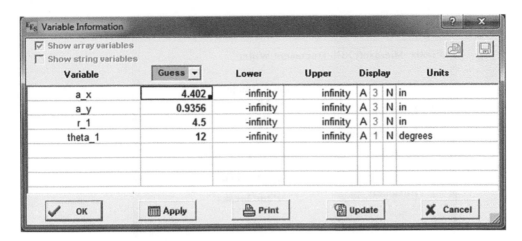

Figure A.6   Variable Information

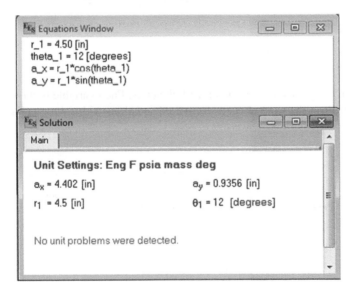

Figure A.7   Equations and solve again

inches as the units for a_x and a_y. You can also set the lower and upper limits for any variable in this window. Pick OK. If nonlinear equations are involved, you would need to set the initial guess for each variable before you solve the problem to ensure convergences.

Since the EES was run once, the guesses for a_x and a_y are the solution from the previous analysis (see Figure A.7).

>**Calculate**>**Solve** again.

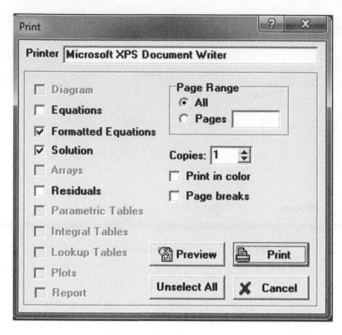

Figure A.8   Print window

This time, there aren't any errors or problems with the units. Each variable is shown with its correct value and units. Print the work sheet as formatted equations with solution (Figure A.8).

>File>Print.
Pick Print.

Pick Print.

File:C:\Users\m-rider.ONU\Desktop\Example.EES   9/8/2012 11:33:05 AM Page 1
EES Ver. 9.183: #531: for use by Engineering at Ohio Northern University

$r_1 = 4.5$ [in]

$\theta_1 = 12$ [degrees]

Solving for the position of point A

$a_x = r_1 \cdot \cos\left[\theta_1\right]$

$a_y = r_1 \cdot \sin\left[\theta_1\right]$

SOLUTION
Unit Settings: Eng F psia mass deg
$a_x = 4.402$ [in] $a_y = 0.9356$ [in] $r_1 = 4.5$ [in]
$\theta_1 = 12$ [degrees]

No unit problems were detected.

Summarize the answer with 3 significant figures and the appropriate units, and then box it below the EES solution.

The position of point A is $(A_x, A_y) = (4.40, 0.936)$ inches.

Example A.2 Vector loop analysis using EES

Given (Figure A.9):

$R_1 = 4.50$ inches at $\theta_1 = 0°$, $R_2 = 1.20$ inches, $R_3 = 4.00$ inches, and $R_4 = 2.10$ inches

Find: θ_3 and θ_4, for $0° \le \theta_2 \le 360°$ in 20 degree increments

Solution:

Vector equation: **R2 + R3 = R1 + R4**

$R_2\cos(\theta_2) + R_3\cos(\theta_3) = R_1\cos(\theta_1) + R_4\cos(\theta_4)$
$R_2\sin(\theta_2) + R_3\sin(\theta_3) = R_1\sin(\theta_1) + R_4\sin(\theta_4)$

Start EES, and pick Continue.

>Options> Unit System, select English, and pick OK (note angles in degrees).

In the Equations window, enter the known variables with units in square brackets and the equations to be solved in any order. Comments are added by beginning a line with a double slash (//) and brackets { } or enclosing the text in double quotes (Figure A.10). Multiple equations can be placed on the same line if they are separated by a semicolon.

>Calculate>Check/Format.

Note that there are 7 equations and 8 variables because we are going to create a table of θ_2 values to satisfy the 8th variable. We will now create the table (see Figure A.11).

>Tables >New Parametric Table.

Select theta_2, and then pick the Add button. Do the same for theta_3 and theta_4. Set the number of runs to 360/20+1 = 19. Enter a name for the table. Pick OK.

Set the first value to zero. Choose **Increment**. Set the increment to **20**. Pick **Apply**. The first column will be filled in with values from 0 to 360 in increments of 20. Pick **OK** (Figure A.12).

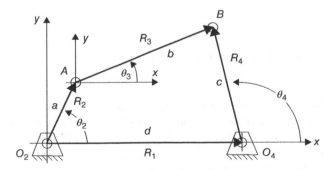

Figure A.9   4-bar linkage

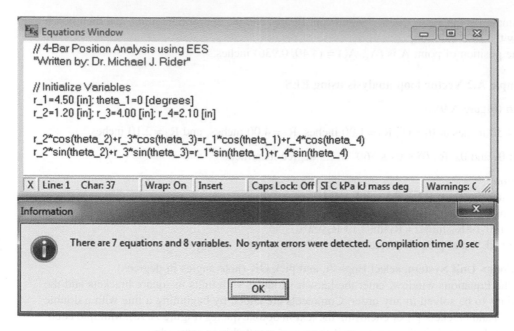

Figure A.10 Equations for table solution

Figure A.11 New Parametric Table window

In the Equations window, right-click to bring up the Variable Info window. Set the lower limit for all links to 0. Set the lower limit for angles to −360 and the upper limit to 360. Set the units for theta_2, theta_3, and theta_4 to degrees. In the fifth column, change the number format from automatic to fixed with 1 decimal place. Set the initial guesses for θ_3 and θ_4 to appropriate values since this is a nonlinear system of equations. Pick **OK** (see Figure A.13).

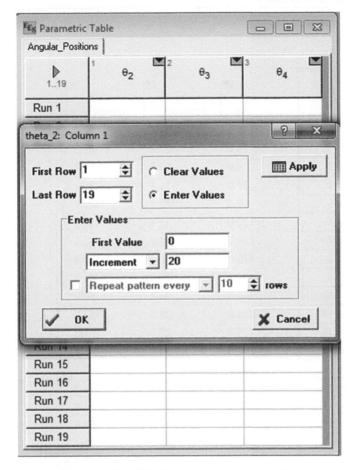

Figure A.12 Fill table for independent variable

Variable	Guess ▾	Lower	Upper	Display			Units	K
r_1	4.5	0.0000E+00	infinity	A	3	N	in	
r_2	1.2	0.0000E+00	infinity	A	3	N	in	
r_3	4	0.0000E+00	infinity	A	3	N	in	
r_4	2.1	0.0000E+00	infinity	A	3	N	in	
theta_1	0.0	-3.6000E+02	3.6000E+02	F	1	N	degrees	
theta_2	0.0	-3.6000E+02	3.6000E+02	F	1	N	degrees	
theta_3	20.0	-3.6000E+02	3.6000E+02	F	1	N	degrees	
theta_4	90.0	-3.6000E+02	3.6000E+02	F	1	N	degrees	

Figure A.13 Variable Information window

EES Parametric Table

Angular_Positions

1.. 19	θ_2 [degrees]	θ_3 [degrees]	θ_4 [degrees]
Run 1	0.0	31.6	87.1
Run 2	20.0	24.7	82.9
Run 3	40.0	19.3	84.7
Run 4	60.0	15.4	91.2
Run 5	80.0	12.7	100.7
Run 6	100.0	11.1	111.9
Run 7	120.0	10.2	123.6
Run 8	140.0	10.3	135.0
Run 9	160.0	11.7	144.5
Run 10	180.0	14.8	150.8
Run 11	200.0	20.0	152.9
Run 12	220.0	26.5	151.2
Run 13	240.0	33.2	146.7
Run 14	260.0	39.2	140.1
Run 15	280.0	43.5	131.5
Run 16	300.0	45.2	121.0
Run 17	320.0	43.6	109.0
Run 18	340.0	38.6	96.7
Run 19	360.0	31.6	87.1

Figure A.14   Table solution

>**Calculate** >**Solve Table**, or in the Parametric Table window, pick the green arrow to solve the system of equations for these 19 cases and fill in the answers for θ_3 and θ_4 (see Figure A.14).

Now, let's plot the results of theta_3 versus theta_2 (see Figure A.15).

>Plots >New Plot window >X–Y plots.

Name the plot that is desired. Select theta_2 for the X-axis variable, and select theta_3 for the Y-axis variable. Set the X-axis maximum to 360 with an interval of 45. Pick OK.

The same could be done for theta_4 versus theta_2 (see Figure A.16). Print the results. Box in any answers with the appropriate units.

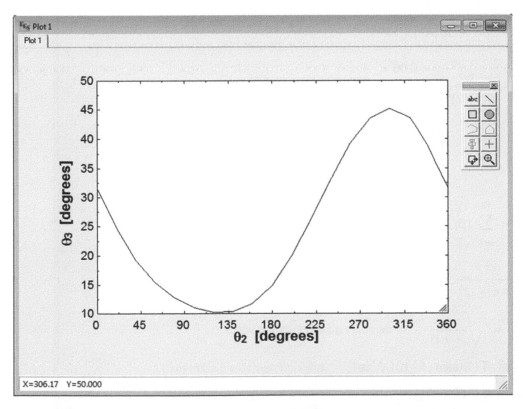

Figure A.15 New Plot window

Figure A.16 Plot window

Formatted Equations Window

The Formatted Equations window displays the equations entered in the Equations window in an easy-to-read mathematical format as shown in the sample in the following text (see Figures A.17 and A.18).

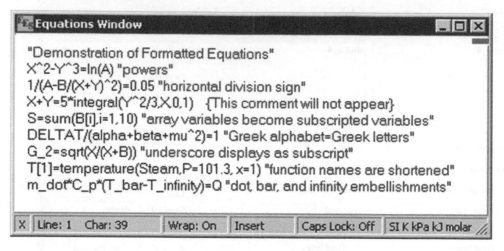

Figure A.17 Equations window

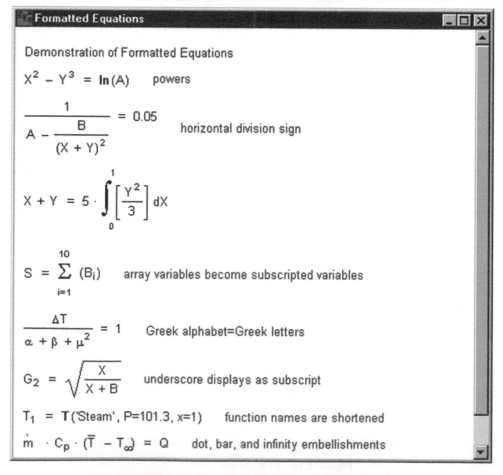

Figure A.18 Formatted Equations window

Comments appearing in quotes in the Equations window are displayed in the Formatted Equations window, but comments in braces are not displayed. An examination of the Formatted Equations window reveals a number of EES features that improve the display. Array variables, such as B[1], are displayed as subscripted variables. Sums and integrals are represented by their mathematical signs. If a variable name contains an underscore, the underscore will signify the beginning of a subscript, as in variable G_2. Placing _dot, _ddot, _bar, _hat, or _tilde after a variable name places a dot, double dot, bar, hat (^), or tilde (~) centered over the name.

Variables having a name from the Greek alphabet are displayed with the equivalent Greek letter. A vertical bar character in a variable name signifies the start of a superscript. For example, theta|° will display as $\theta°$. A special form is provided for variables beginning with DELTA. For example, DELTAtheta displays as $\Delta\theta$. Also, deltatheta displays as $\delta\theta$.

Some Common Built-in Functions

EES has a large library of built-in mathematical functions. Many of these (e.g., sin, cos, tan, arctan, etc.) are particularly useful for engineering applications. EES provides a function to convert units and functions that help manipulate complex numbers.

All functions are case insensitive; thus, COS(x), Cos(x), and cos(x) are the same function.

abs(X)	returns the absolute value of the argument. In complex mode, **abs** returns the magnitude of the complex argument (see also **magnitude**(X)).
angle(X), **angleDeg**(X), and **angleRad**(X)	all return the angle (also called amplitude or argument) of complex variable X. If X = X_r + i*X_i, this function returns arctan(X_i/X_r). The function *returns the angle in the correct quadrant* of the complex plane. Note that the Angle functions are used to extract the angle of a complex number variable or expression.
Angle(X)	will return the angle in either degrees or radians depending on the Trig Function setting in the Unit System dialog. The value returned will be between −180 degrees and +180 degrees or $-\pi$ radians and $+\pi$ radians.
angleDeg(X)	will always return the angle in degrees between −180 and +180.
angleRad(X)	will always return the angle in radians between $-\pi$ and $+\pi$.
arcCos(X)	returns the angle which has a cosine equal to the value of the argument. The units of the angle (degrees or radians) will depend on the unit choice made for trigonometric functions with the Unit System command. The value returned will be between 0 degrees and +180 degrees or 0 radians and $+\pi$ radians.
arcSin(X)	returns the angle which has a sine equal to the value of the argument. The units of the angle (degrees or radians) will depend on the unit choice made for trigonometric functions with the Unit System command. The value returned will be between −90 degrees and +90 degrees or $-\pi/2$ radians and $+\pi/2$ radians.

arcTan(X)   returns the angle which has a tangent equal to the value of the argument. The units of the angle (degrees or radians) will depend on the unit choice made for trigonometric functions with the Unit System command. The value returned will be between −90 degrees and +90 degrees or $-\pi/2$ radians and $+\pi/2$ radians.

cis(X)   returns a complex number that is equal to cos(X)+i*sin(X). The units of the angle (degrees or radians) will depend on the unit choice made for trigonometric functions with the Unit System command. For example, Y=3*cis (30deg) will define a complex number with a magnitude of 3 and an angle of 30° regardless of the unit setting.

However, you can ensure that the angle you enter is in degrees or radians regardless of the unit system setting by appending deg or rad to the number (with no spaces). For example, the value of Y can be set to the same complex constant in any one of the following four ways:

$$Y = 2.598 + 1.500 * i$$
$$Y = 3.000*cis(30deg)$$
$$Y = 3.000 < 30deg$$
$$Y = 3.000 <0.5236rad$$

conj(X)   returns the complex conjugate of a complex variable X. If $X = X\_r + i*X\_i$, this function returns $X\_r - i*X\_i$. Note that this function returns a complex result. The EES equation, Y=conj(X), will set the real part of Y (Y_r) to the real part of X and the imaginary part of Y to negative of the imaginary part (Y_i).

convert('From', 'To')   returns the conversion factor needed to convert units from the unit designation specified in the 'From' string to that specified in the 'To' string. The single quote marks are optional. For example, F2I = **convert**(ft^2, in^2) will set F2I to a value of 144 because 1 square foot is 144 square inches. Combinations of units and multiple unit terms may be entered.

cos(X)   will return the cosine of the angle provided as the argument. The required units (degrees or radians) of the angle are controlled by the unit choice made for trigonometric functions with the Unit System command.

exp(X)   will return the value "e = 2.718 ..." raised to the power of the argument X.

if (A, B, X, Y, Z)   allows conditional assignment statements in the Equations window. If A<B, the function will return a value equal to the value supplied for X; if A=B, the function will return the value of Y; and if A>B, the function will return the value of Z. In some problems, use of the **if** function may cause numerical oscillation. It is preferable to use the *If-Then-Else*, *Repeat-Until*, and *GoTo* statements in a function or procedure for conditional assignments.

imag(X)   returns the imaginary part of a complex variable X. If $X = X\_r + i*X\_i$, this function returns X_i. Note that X = 4*i will set X to 0 + 4*i. If you wish to only set the imaginary part of X, you can enter X_i = 4.0.

ln(X)   will return the natural logarithm of the argument.

log10(X)   will return the base 10 logarithm of the argument.

magnitude(X)   returns the magnitude (also called modulus or absolute value) of a complex variable X. In complex mode, the **abs** function also returns the magnitude. If $X = X_r + i*X_i$, this function returns sqrt(X_r^2+X_i^2).

pi   is a reserved variable name which has the value of 3.1415927.

real(X)   returns the real part of a complex variable X. If $X = X_r + i*X_i$, this function returns X_r. Note that X = 4 will set X to 4 + i*0. If you wish to only set the real part of variable X, you can enter X_r = 4.

round(X)   will return a value equal to the nearest integer value of the argument.

sign(X)   will return the sign (+1 or −1) of the argument.

sin(X)   will return the sine of the angle provided as the argument. The required units (degrees or radians) of the angle are controlled by the unit choice made for trigonometric functions with the Unit System command.

sqrt(X)   will return the square root of the value provided as the argument which must be greater than or equal to zero.

tan(X)   will return the tangent of the angle provided as the argument. The required units (degrees or radians) of the angle are controlled by the unit choice made for trigonometric functions with the Unit System command.

$Complex   ON/OFF i/j turns on or off complex number mode and defines the symbol "i" or "j" as the square root of minus one. Like other directives, $CHECKUNITS should be on a line by itself.

$CheckUnits   ON/OFF and AutoOn/AutoOff controls whether the unit checking algorithms will be applied to the equations that follow. Like other directives, $CHECKUNITS should be on a line by itself. By default, unit checking is ON for all equations. When EES encounters a $CHECKUNITS Off directive, equations that follow will not be subject to unit checking. $CHECKUNITS ON restores the unit checking to the equations that follow.

Single-line If-Then-Else Statements

EES functions and procedures support several types of conditional statements. These conditional statements cannot be used in modules or in the main body of an EES program. The most common conditional is the *If-Then-Else* statement. Both single-line and multiple-line formats are allowed for *If-Then-Else* statements. The single-line format has the following form.

```
If (Conditional Test) Then Statement 1 Else Statement 2
```

The conditional test yields a *true* or *false* result. The format is very similar to that used in Pascal. Recognized operators are =, <, >, <=, >=, and <> (for not equal). The parentheses around the conditional test are optional. The *Then* keyword and Statement 1 are required. Statement 1 can be either an assignment or a *GoTo* statement. The *Else* keyword and Statement 2 are optional. In the single-line format, the entire *If-Then-Else* statement must be placed on one line with 255 or fewer characters. The following example function uses *If-Then-Else* statements:

```
If (x<y) Then m:=x Else m:=y
If (m>z) Then m:=z
```

The AND and OR logical operators can also be used in the conditional test of an *If-Then-Else* statement. EES processes the logical operations from left to right unless parentheses are supplied to change the parsing order. Note that the parentheses around the $(x>0)$ and $(y<>3)$ are required in the following example to override the left to right logical processing and produce the desired logical effect.

```
If (x>y) or ((x<0) and (y<>3)) Then z:=x/y Else z:=x
```

Multiple-line If-Then-Else Statements

The multiple-line *If-Then-Else* statement allows a group of statements to be executed conditionally. This conditional statement can be used in functions and procedures, but not in modules or the main body of an EES program. The format is as follows:

```
If (Conditional Test) Then
Statement
Statement
...
Else
Statement
Statement
...
EndIf
```

The *If* keyword, the conditional test, and *Then* keyword must be on the same line. The parentheses around the conditional test are optional. The statements which are to be executed if the conditional test is true appear on following lines. These statements may include additional *If-Then-Else* statements so as to have nested conditionals. An *Else* (or *EndIf*) keyword terminates this first group of statements. The *Else* keyword should appear on a line by itself, followed by the statements which execute if the conditional test is false. The *EndIf* keyword, which terminates the multiple-line *If-Then-Else* statement, is required, and it must appear on a line by itself.

Appendix B

MATLAB

Top of form

MATLAB is an acronym for "Matrix Laboratory." MATLAB is a high-performance computer language for technical applications. It combines computations with visualization and easy-to-use programming. MATLAB is an interactive system whose basic data element is an array that does not require dimensioning. A scalar variable is simply an array of one element.

The help window is interactive and can be used to obtain information on any feature of MATLAB. This window can be opened from the help menu in the toolbar of any MATLAB window.

Starting MATLAB

The default installation program will create a directory named C:\Program Files \MATLAB in which the MATLAB files are placed. The MATLAB icon shown above will identify the program. Double-clicking the left mouse button on the icon will start the program.

MATLAB has 3 tabs as shown in Figure B.1: HOME, PLOTS, and APPS. Each has a toolbar below the tab for commonly used commands in that area. If you rest the cursor over a toolbar icon and wait a few seconds, a few words will appear to explain the function of that icon.

Design and Analysis of Mechanisms: A Planar Approach, First Edition. Michael J. Rider.
© 2015 John Wiley & Sons, Ltd. Published 2015 by John Wiley & Sons, Ltd.

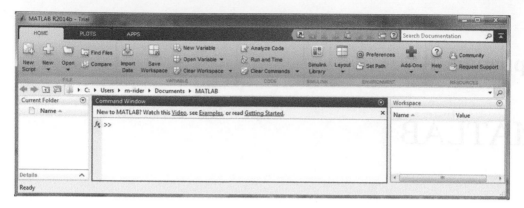

Figure B.1   MATLAB windows and tabs

Some Common MATLAB Programming Commands or Symbols

=   Assignment operator (equal sign)
          The assignment operator assigns a value to a variable. The general form is:

                      Variable = numerical value or computational expression

;   Semicolon
          When the semicolon is added to the end of an equation, it disables the display of the
          answer to that equation. Without the semicolon at the end of the expression, the result is
          displayed in the display window

%   Percent symbol
          When the symbol % (percent) is typed at the beginning of a line, the line is designated
          as a comment. This line has no effect on the execution of the program. The % character
          followed by some text makes the rest of the line a comment

input   Prompt for user input
          RESULT = input(PROMPT) displays the PROMPT string on the screen, waits for
          input from the keyboard, evaluates any expressions in the input, and returns the
          value in RESULT. To evaluate expressions, input accesses variables in the current
          workspace. If you press the return key without entering anything, input returns an
          empty matrix

[ ]   Vector created
          A vector is created by typing the elements or numbers inside square brackets. To create a
          row vector, type the elements with a space or a comma between the elements inside the
          square brackets. To create a column vector, type the square bracket and then the elements
          with a semicolon separating each new row followed by a closing square bracket

for   Repeat statements a specific number of times
          The columns of the expression are stored one at a time in the variable, and then the
          following statements, up to the END, are executed. If the increment is left out, it
          defaults to one. The general form is:

```
for variable = first:increment:last
    statement
    …
    statement
END
```

while   Repeat statements an indefinite number of times
         The statements are executed while the real part of the expression has all nonzero
         elements. The expression is usually the result of (expr rop expr) where rop is ==, <, >,
         <=, >=, or ~=. The general form of a while statement is:

```
while expression
    statements
END
```

if   Conditionally execute statements. It must end with an END statement
         Upper- and lowercase are treated as the same as far as elseif, else, and end are
         concerned
         The general form of the if statement is:

```
if expression
    statements
ELSEIF expression
    statements
ELSE
    statements
END
```

The statements are executed if the real part of the expression has all nonzero elements.
         The ELSE and ELSEIF parts are optional. Zero or more ELSEIF parts can be used
as well as nested if's. The expression is usually of the form expr rop expr where the
relational operator, rop, is ==, <, >, <=, >=, or ~=
         Example:

```
if a == b
    x(a,b) = 2;
elseif abs(a-b) == 1
    x(a,b) = -1;
else
    x(a,b) = 0;
end
```

Some Common MATLAB Functions

abs   Absolute value
        abs(X) is the absolute value of the elements of X. When X is complex, abs(X) is the
        complex modulus (magnitude) of the elements of X

angle   Phase angle
            angle(H) returns the phase angles, in radians, of a variable or a matrix with
            complex elements. The angle returned is between $-\pi$ and π

acos   Inverse cosine, result in radians
            acos(X) is the arccosine of the elements of X. The angle returned is between 0 and
            π. Complex results are obtained if ABS(x) >1.0 for some element

acosd   Inverse cosine, result in degrees
            acosd(X) is the inverse cosine, expressed in degrees, of the elements of X.
            The angle returned is between 0 and 180

asin   Inverse sine, result in radians
            asin(X) is the arcsine of the elements of X. The angle returned is between $-\pi/2$
            and $\pi/2$. Complex results are obtained if ABS(x) >1.0 for some element

asind   Inverse sine, result in degrees
            asind(X) is the inverse sine, expressed in degrees, of the elements of X. The angle
            returned is between -90 and 90

atan   Inverse tangent, result in radians
            atan(X) is the arctangent of the elements of X. The angle returned is between $-\pi/2$
            and $\pi/2$

atand   Inverse tangent, result in degrees
            atan(X) is the arctangent of the elements of X in degrees. The angle returned is
            between -90 and 90

atan2(y,x) Four-quadrant inverse tangent (in radians)
            P = atan2(Y, X) returns an array P the same size as X and Y containing the
            element-by-element, four-quadrant inverse tangent (arctangent) of Y and X,
            which must be real. The angle returned is between $-\pi$ and π
                Elements of P lie in the closed interval [−pi, pi], where pi is the MATLAB®
            floating point representation of π. Atan2 uses sign(Y) and sign(X) to determine
            the specific quadrant (see Figure B.2)

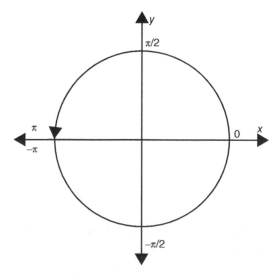

Figure B.2   Range of atan2

For any complex number, $z = x + i\,y$, it is converted to polar coordinates using the functions *abs* and *atan2*:

```
r = abs(z)
theta = atan2(imag(z), real(z))
```

Example:

```
V = -5 + 12i;
M = abs(V)
theta = atan2(imag(V), real(V)) = angle(V)

M = 13
theta = 1.9656
```

To convert it back to the original complex number, V2:

```
V2 = M * exp(i * theta)
V2 = -5.0000 +12.0000i
```

atan2d   Four-quadrant inverse tangent (in degrees)
(y,x) D = atan2d(Y, X) returns the four-quadrant inverse tangent of points specified in the x–y plane. The result, D, is expressed in degrees. The angle returned is between −180 and 180

Elements of D lie in the closed interval [−180, 180] (see Figure B.3). Atan2d uses sign(Y) and sign(X) to determine the specific quadrant

Example:

```
V = -5 -12i
M = abs(V)
thetadeg = atan2d(imag(V), real(V))

M = 13
thetadeg = -112.6199
```

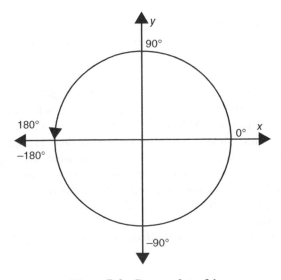

Figure B.3   Range of atan2d

To convert it back to the original complex number, V2:

```
V2 = M * exp(i * thetadeg * pi/180)
V2 = -5.0000 -12.0000i
```

complex   Create complex array
C = complex(A,B) returns the complex result A + Bi, where A and B are identically sized real N–D arrays, matrices, or scalars of the same data type. Note that in the event that B is all zeros, C is complex with all zero imaginary part, unlike the result of the addition A+0i, which returns a strictly real result

    C = complex(A) for real A returns the complex result C with real part A and all zero imaginary part. Even though its imaginary part is all zero, C is complex and so isreal(C) returns false. If A is complex, C is identical to A

    The complex function provides a useful substitute for expressions such as A+1i*B or A+1j*B in cases when A and B are not single or double or when B is all zero

conj   Complex conjugate
conj(X) is the complex conjugate of X. For a complex X, conj(X) = REAL(X) – i*IMAG(X)

cos   Cosine of argument in radians
cos(X) is the cosine of the elements of X

cosd   Cosine of argument in degrees
cosd(X) is the cosine of the elements of X, expressed in degrees. For odd integers n, cosd(n*90) is exactly zero, whereas cos(n*pi/2) reflects the accuracy of the floating point value for pi

exp   Exponential
exp(X) is the exponential of the elements of X, e to the X
For complex Z = X+i*Y, exp(Z) = exp(X)*(COS(Y)+i*SIN(Y)). For complex Z, X is the magnitude and Y is the angle in radians

expm1   Compute EXP(X)–1 accurately
expm1(X) computes EXP(X)–1, compensating for the roundoff in EXP(X). For small real X, expm1(X) should be approximately X, whereas the computed value of EXP(X)–1 can be zero or have high relative error

format   Set output format
format with no inputs sets the output format to the default appropriate for the class of the variable. For float variables, the default is format SHORT
format does not affect how MATLAB computations are done. Computations on float variables, namely, single or double, are done in appropriate floating point precision, no matter how those variables are displayed. Computations on integer variables are done natively in integer. Integer variables are always displayed to the appropriate number of digits for the class, for example, 3 digits to display the INT8 range—128:127
format SHORT and LONG do not affect the display of integer variables

format may be used to switch between different output display formats of all float variables as follows:

format SHORT: Scaled fixed point format with 5 digits
format LONG: Scaled fixed point format with 15 digits for double and 7 digits for single
format SHORTE: Floating point format with 5 digits
format LONGE: Floating point format with 15 digits for double and 7 digits for single
format SHORTG: Best of fixed or floating point format with 5 digits
format LONGG: Best of fixed or floating point format with 15 digits for double and 7 digits for single
format SHORTENG: Engineering format that has at least 5 digits and a power that is a multiple of three
format LONGENG: Engineering format that has exactly 16 significant digits and a power that is a multiple of three

format may be used to switch between different output display formats of all numeric variables as follows:

format HEX: Hexadecimal format
format +:   The symbols +, −, and blank are printed for positive, negative, and zero elements. Imaginary parts are ignored
format BANK: Fixed format for dollars and cents
format RAT: Approximation by ratio of small integers. Numbers with a large numerator or large denominator are replaced by *

format may be used to affect the spacing in the display of all variables as follows:

format COMPACT: Suppresses extra line-feeds
format LOOSE: Puts the extra line-feeds back in

imag   Complex variable's imaginary part
$imag(X)$ is the imaginary part of X. Use i or j to enter the imaginary part of complex numbers

log   Natural logarithm
$log(X)$ is the natural logarithm of the elements of X. Complex results are produced if X is not positive

log1p   Compute $LOG(1+X)$ accurately
$log1p(X)$ computes $LOG(1+X)$, without computing $1+X$ for small X. Complex results are produced if $X < -1$. For small real X, $log1p(X)$ should be approximately X, whereas the computed value of $LOG(1+X)$ can be zero or have high relative error

log10   Common (base 10) logarithm
$log10(X)$ is the base 10 logarithm of the elements of X. Complex results are produced if X is not positive

pi   3.1415926535897…
The constant can be accessed by referencing the variable pi or the function pi()
$pi = 4*atan(1) = imag(log(-1)) = 3.1415926535897…$

real   Complex variable's real part
                 real(X) is the real part of X. Use i or j to enter the imaginary part of complex
                 numbers

isreal   True for real numbers or arrays
                 isreal(X) returns 1 if X does not have an imaginary part and 0 otherwise

round   Rounds toward nearest decimal or integer
                 round(X) rounds each element of X to the nearest integer
                 round(X, N), for positive integers N, rounds to N digits to the right of the decimal
                     point. If N is zero, X is rounded to the nearest integer
                 If N is less than zero, X is rounded to the left of the decimal point
                 N must be a scalar integer
                 round(X, N, 'significant') rounds each element to its N most significant digits,
                     counting from the most significant or left side of the number
                 N must be a positive integer scalar
                 round(X, N, 'decimals') is equivalent to round(X, N)
                 For complex X, the imaginary and real parts are rounded independently

sign   Signum function
                 For each element of X, sign(X) returns 1 if the element is greater than zero, 0 if
                 it equals zero, and −1 if it is less than zero. For the nonzero elements of complex
                 X, sign(X) = X ./ ABS(X)

sin   Sine of argument in radians
                 sin(X) is the sine of the elements of X

sind   Sine of argument in degrees
                 sind(X) is the sine of the elements of X, expressed in degrees. For integers n,
                 sind(n*180) is exactly zero, whereas sin(n*pi) reflects the accuracy of the
                 floating point value of pi

sqrt   Square root
                 sqrt(X) is the square root of the elements of X. Complex results are produced if
                 X is not positive

realsqrt   Real square root or error
                 realsqrt(X) is the square root of the elements of X. An error is produced if X is
                 negative

tan   Tangent of argument in radians
                 tan(X) is the tangent of the elements of X in radians

tand   Tangent of argument in degrees
                 tand(X) is the tangent of the elements of X, expressed in degrees. For odd
                 integers n, tand(n*90) is infinite, whereas tan(n*pi/2) is large but finite, reflecting
                 the accuracy of the floating point value of pi

Some Common MATLAB 2-D Graphing Procedures

plot   Linear plot
             plot(X,Y) plots vector Y versus vector X. If X or Y is a matrix, then the vector is
                 plotted versus the rows or columns of the matrix, whichever line up. If X is a scalar

and Y is a vector, disconnected line objects are created and plotted as discrete points vertically at X

plot(Y) plots the columns of Y versus their index. If Y is complex, plot(Y) is equivalent to plot(real(Y),imag(Y)). In all other uses of plot, the imaginary part is ignored

plot(X1,Y1,X2,Y2,X3,Y3,...) combines the plots defined by the (X,Y) pairs, where the X's and Y's are vectors or matrices

axis   Control axis scaling and appearance

axis([XMIN XMAX YMIN YMAX]) sets scaling for the x-axis and y-axis on the current plot

axis EQUAL sets the aspect ratio so that equal tick mark increments on the x-axis and y-axis equal in size. This makes a circle look like a circle, instead of an ellipse

axis SQUARE makes the current axis box square in size

axis NORMAL restores the current axis box to full size and removes any restrictions on the scaling of the units

zoom   Zoom in and out on a 2-D plot

zoom ON turns zoom on for the current figure

zoom OFF turns zoom off in the current figure

grid   Grid lines

grid ON adds major grid lines to the current axes

grid OFF removes major and minor grid lines from the current axes

grid MINOR toggles the minor grid lines of the current axes

hold   Hold current graph

hold ON holds the current plot and all axis properties, including the current color and line style, so that subsequent graphing commands add to the existing graph without resetting the color and line style

hold OFF returns to the default mode whereby PLOT commands erase the previous plots and reset all axis properties before drawing new plots

title   Graph title

title('text') adds text at the top of the current axis

xlabel   x-axis label

xlabel('text') adds text beside the x-axis on the current axis

ylabel   y-axis label

ylabel('text') adds text beside the y-axis on the current axis

text   Text annotation

text(X,Y,'string') adds the text in the quotes to location (X,Y) on the current axes, where (X,Y) is in units from the current plot. If X and Y are vectors, text writes the text at all locations given. If 'string' is an array the same number of rows as the length of X and Y, text marks each point with the corresponding row of the 'string' array

gtext   Place text with mouse

gtext('string') displays the graph window, puts up a crosshair, and waits for a mouse button or keyboard key to be pressed. The crosshair can be positioned with the mouse (or with the arrow keys on some computers). Pressing a mouse button or any key writes the text string onto the graph at the selected location

Optimization Toolkit for MATLAB

If your version of MATLAB contains the optimization toolkit, then the following function is available to you. It can be used to solve simultaneous nonlinear equations such as those derived using vector loops. The student version of MATLAB does not contain this toolkit.

fsolve   Solve a system of nonlinear equations
fsolve(funct,guess) starts at values of guess and tries to solve the set of equations defined in the function funct. The function funct may be a single function of multiple variables or a system of equations defined in a column matrix. Each equation must be set equal to zero. The starting point guess is a row matrix containing the initial guesses. An example follows (see Figure B.4):

```
function twoeqns
guess34 = [20, 75];  %Theta3 = 20 degrees and Theta 4 =
75 degrees
theta34 = fsolve (@eqns, guess34)

function theta = eqns(x)
th3=x(1);
th4=x(2);
L1=10;
th1=8;
L2=4;
th2=60;
L3=10;
L4=7;
theta = [L2*sin(th2) + L3*sin(th3) - L4*sin(th4) -
        L1*sin(th1);
        L2*cos(th2) + L3*cos(th3) - L4*cos(th4) -
        L1*cos(th1)];
```

$$L_2 \sin\theta_2 + L_3\sin\theta_3 - L_4\sin\theta_4 - L_1\sin\theta_1 = 0$$
$$L_2\cos\theta_2 + L_3\cos\theta_3 - L_4\cos\theta_4 - L_1\cos\theta_1 = 0$$

>> twoeqns   % To run the MATLAB code and solve for the two unknown angles

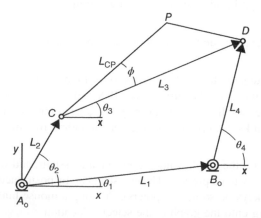

Figure B.4   4-bar linkage with vectors

Further Reading

Here is a list of References I used over the past 37 years along with my personal notes that were used in conjunction with the writing of this textbook. M.J. Rider

Erdman, Arthur G., Sandor, George N., and Kota, Sridhar, *Mechanism Design, Analysis and Synthesis*, Volume 1, 4th ed., Prentice Hall, Inc., 2001.

Hall, Allen S. Jr., *Kinematics and Linkage Design*, Balt Publishers, 1966.

Hrones, John A., and Nelson, George L., *Analysis of the Four-Bar Linkage*. New York: The Technology Press of MIT and John Wiley & Sons, Inc., 1951.

Kimbrell, Jack T., *Kinematics Analysis and Synthesis*, 1st ed., McGraw-Hill, Inc., 1991.

Lynwander, Peter, *Gear Drive Systems, Design and Application*, 1st ed., Marcel Dekker, Inc., 1983.

Mabie, Hamilton H., and Reinholtz, Charles F., *Mechanisms and Dynamics of Machinery*, 4th ed., John Wiley & Sons, Inc., 1987.

Martin, George H., *Kinematics and Dynamics of Machines*, 2nd ed., Waveland Press, Inc. 1982.

Mott, Robert L., *Machine Elements in Mechanical Design*, 3rd ed., Prentice Hall, Inc., 1999.

Norton, Robert L., *Design of Machinery*, 3rd ed., McGraw-Hill Companies, Inc., 2004.

Selby, Samuel M., *Standard Mathematical Tables*, 16th ed., The Chemical Rubber Company, 1968.

Uicker Jr., John J., Pennock, Gordon R., and Shigley, Joseph E., *Theory of Machines and Mechanisms*, 3rd ed., Oxford University Press, 2003.

Waldron, Kenneth J., and Kinel, Gary L., *Kinematics, Dynamics, and Design of Machinery*, 2nd ed., John Wiley & Sons, Inc., 2004.

Wilson, Charles E., and Sadler, J. Peter, *Kinematics and Dynamics of Machinery*, 2nd ed., HarperCollins College Publishers, 1993.

Design and Analysis of Mechanisms: A Planar Approach, First Edition. Michael J. Rider.
© 2015 John Wiley & Sons, Ltd. Published 2015 by John Wiley & Sons, Ltd.

Index

Design and Analysis of Mechanisms: A Planar Approach, First Edition. Michael J. Rider.
© 2015 John Wiley & Sons, Ltd. Published 2015 by John Wiley & Sons, Ltd.

Printed and bound by CPI Group (UK) Ltd, Croydon, CR0 4YY

27/10/2024

14580312-0004